SANTA A
D0382576

THE CAT OWNER'S PROBLEM SOLVER

A READER'S DIGEST BOOK

Published by The Reader's Digest
Association Limited
11 Westferry Circus
Canary Wharf
London E14 4HE

Reprinted 1999 (twice)

ISBN 0 276 42356 9

A Cataloguing-in-Publication record for this book is available from the British Library

Conceived, designed and produced by
Andromeda Oxford Limited

Project Editor:	*Susan Kennedy*
Editor:	*Lauren Bourque*
Art Director:	*Chris Munday*
Senior Designers:	*Martin Anderson*
	Frankie Wood
Picture Manager:	*Claire Turner*
Picture Researcher:	*Valerie Mulcahy*
Proofreader:	*Lynne Elson*
Indexer:	*Ann Barrett*
Production Director:	*Clive Sparling*
Production Assistant:	*Nicolette Colborne*

The acknowledgments that appear on page 202 are hereby made a part of this copyright page.

Printed in Slovakia by Polygraf Print, Prešov.

THE CAT OWNER'S PROBLEM SOLVER

JOHN AND CAROLINE BOWER

PUBLISHED BY THE READER'S DIGEST ASSOCIATION LIMITED
LONDON • NEW YORK • SYDNEY • CAPE TOWN • MONTREAL

INTRODUCTION

MORE PEOPLE THAN EVER OWN A CAT TODAY. ALTHOUGH SOME CATS ARE HAPPIEST LIVING *outdoors, there are many breeds that make ideal pets for those living in flats in cities. In most cases, the relationship is a happy one, but as veterinarians working in a busy practice, and as cat owners ourselves, we know that things can sometimes go wrong. From the questions that owners ask us, we are aware of what aspects of care concern them most, from choosing a young kitten, feeding a correct diet, grooming a longhaired cat, to caring for it in old age.*

Behavioural and health problems are best dealt with by preventing them from arising in the first place. By helping owners understand the way their cat thinks, behaves and communicates, we hope to make it easier to eradicate unpleasant habits such as clawing or spraying indoors. We also give practical help on such situations as owning a cat and having small children.

The section on Your Cat's Health offers advice on preventing disease, neutering, mating and raising kittens. We identify the illnesses and physical conditions that owners are most likely to encounter, with information on symptoms and treatment. Finally, we provide information on some of the most popular cat breeds, describing their particular characteristics as well as some potential problems.

Colleagues at The Veterinary Hospital, Plymouth, have helped us in the writing of the book, and their contributions are listed on page 202. We are also grateful to our patients and their owners – without them, we would not have been able to pose the questions and answers that are a feature of the book. *Our aim has been to offer a readable, easy to use, and entertaining guide that will give you many years of untroubled companionship with your chosen pet.*

JOHN & CAROLINE BOWER

CONTENTS

Caring for Your Cat

THERE IS A COMMON MISCONCEPTION THAT cats are ideal companions for people who like the idea of owning a pet but don't have the time to care for one. Nothing could be further from the truth. Although cats may be less demanding than other pets, they still deserve the same degree of commitment from us. The reward of caring for a cat is the presence in your home of one of nature's most beautiful, graceful animals, healthy and contented.

Before taking home a cat, you should be aware of its basic physical requirements. A balanced diet and a supply of fresh, clean water are essential for its health. Your cat will need access to the outside or to a litter tray that is cleaned regularly. Have it vaccinated every year against infectious diseases, and treat it regularly for worms and the inevitable fleas. Be prepared to make additional trips to the vet whenever you suspect there is a health problem. Grooming is always a good idea and is especially necessary for longhaired cats.

Finally, cats need human company and affection. While they tend to be independent, they do enjoy contact with people and may 'adopt' your neighbours if they fail to receive enough attention at home. If you are away from home for a day or more, you must arrange for someone to come in to care for your cat or find a suitable boarding facility.

You and Your Cat

IF YOU HAVE DECIDED TO GET A CAT, YOU ARE in very good company. In this century cats have decisively overtaken dogs as the preferred pet in developed countries, where pet ownership is as widespread in urban areas as in the countryside.

In many ways a cat is the ideal pet for modern living. It is clean, relatively quiet, takes up little space, does not need to be taken out at regular intervals, and is not expensive to maintain. Even people who are not cat lovers concede that they are among nature's most beautiful creatures.

Your Cat's Basic Needs

- ✓ A balanced diet and fresh water available at all times.
- ✓ Access to the outside or a litter tray that is cleaned every day.
- ✓ Regular deworming and antiflea treatment, particularly in summer.
- ✓ Regular vaccinations, health checks and veterinary care if it is sick or injured.
- ✓ Neutering of both males and females unless you intend to use them for breeding.
- ✓ Grooming – essential for longhaired breeds but good for all cats.
- ✓ An identification tag or tattoo so that it can be returned if it wanders away from home.
- ✓ Human companionship and affection. Toys to play with when it is left on its own.

A Love-Hate Relationship

Like most of their larger relatives in the feline family, domestic cats have a long history of being both feared and admired by humans. Their grace and elegance have fascinated many cultures, but their great popularity in ancient Egypt and other cultures was originally due to their ability as hunters of vermin. It was in Egypt that cats were elevated to be the objects of cult worship. Thousands of years later, after the Christian church had become established in Europe, cats came to be strongly linked with pagan beliefs. Despite the fact that cats were associated with several early Christian saints, among them St Jerome and St Agatha, they were often badly treated. They later came to be regarded as symbols of witches and witchcraft, and were objects of suspicion, rumour and persecution.

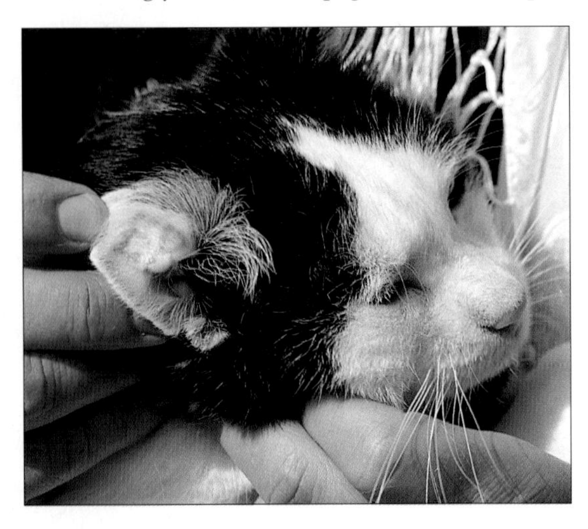

▲ *Your cat should have some kind of identification. A tattoo, as shown here, or a microchip is a permanent marker of its identity and your home address.*

A lingering trace of this can be seen in the persistent prejudice against cats for their occasional aloofness, perceived as 'selfishness'. Cats are no more selfish than any other animal; none of them can be judged fairly by human standards (see the Behaviour section, especially pages 40–47).

A Kitten for Life

Domestic animals never really grow up: no matter how old they are, they remain dependent on their owners. Your cat will probably live for 14 or 15 years, perhaps longer, but in spite of stiff joints and other geriatric complaints, it remains a kitten, looking to you as it did to its mother to provide food, warmth and safety. In addition to these basic ingredients for survival, you should do all you can to ensure the health and comfort

● ***We already have a family dog, but we'd really like to get a cat. Is this a bad idea?***

Not necessarily. It should be easy if your dog is well trained and if you choose a kitten, preferably one that met a dog or dogs before it was 7 weeks old. If you want an adult, it is even more important to find one used to dogs (see pages 48–49).

● ***My son is allergic to cats, but he still wants to get a kitten. How much trouble will it be?***

This depends on how allergic your son is. If he has only a mild allergy, he may be able to live comfortably with a shorthaired cat that sheds very little, such as a Rex, as long as the house is kept well vacuumed and the cat is not allowed into his room. If he has a severe allergy, it is better to choose another pet for him. You might consider a guinea pig or a non-shedding small dog such as a poodle or a Bichon Frise.

● ***I'd like to get a cat for my elderly mother, but if I get a kitten, it may outlive her; and if I get an older cat and it dies, she'll be distraught. Any advice?***

Does your mother want a cat, and is she active enough to care for it and take it to the vet? If so, perhaps you should look for a young adult cat and consider what alternative homes you could place it in if necessary.

of your cat by having it vaccinated against infectious diseases, protecting it against infestation by common parasites, and taking it to your vet promptly if it appears unwell in any way. Neutering is advisable unless you own a pedigree cat that you intend to breed (see pages 74–77).

Although these requirements are comparatively minimal, they nevertheless constitute a serious responsibility that should not be taken lightly. If you attend to your cat's physical needs and provide it with plenty of love and attention, you will be rewarded with a healthy, beautiful, affectionate companion that radiates well-being from its whiskers to the tip of its tail.

▼ *Cats are beautiful animals and a very practical choice of pet for most people's circumstances. Children love them, and people who live alone may find a cat an especially rewarding companion.*

Adopting a Homeless Cat

IF THE BOUNDLESS ENERGY AND MISCHIEF OF A kitten does not appeal to you, consider adopting a homeless cat. It can be enormously satisfying to provide a good home for an abandoned cat, and you will have plenty to choose from. Some may have been given up owing to changed family circumstances or because they had health or behaviour problems; others may be strays or even feral cats (domesticated cats that have been living wild). If you want a particular breed, you may still be able to find it in a shelter, though the animal is unlikely to be of show quality. Some larger breed associations maintain their own shelters for pedigree cats.

A pet should never be chosen for looks alone. Think about what you want and then look for those qualities in the cats you see. This should help to minimise the urge to take home every cat you see at the shelter. Begin by making a list of all the factors that are important in relation to your family and lifestyle. If you live in a flat it is no good bringing home a farm cat that is used to spending all its time hunting.

Next discuss your preferences with someone at the shelter. For example, if you already have a cat at home, be sure to choose one that will happily coexist with another feline: staff at the shelter will have observed which cats do not tolerate others. Information may have been provided by the former owners, but it may not be reliable, especially if the cat has a specific problem. It is to your advantage to find out as much as possible. Remember that no cat will be at its best if it is sick or stressed.

Choosing a Healthy Cat

A healthy cat should have clear, bright eyes, a glossy coat, good body condition, and show no symptoms such as lameness, vomiting, or diarrhoea, a discharge from the nose or eyes, or laboured breathing. A rescued cat is unlikely to be in peak condition, but if you like the cat and have time and resources to devote to bringing it

● ***We have fallen in love with a magnificent cat at the local shelter, but the manager told us that she's feral. What should we do?***

If you want a cuddly family pet, choose a different cat. Feral cats have not been socialised to humans or other pets, and no amount of time and attention will ever make them into lap cats, or even indoor cats. If this suits you, and you have no young children to plague her, then go ahead and choose her, as long as you are satisfied she is completely healthy. She may never lose her distrust of strangers, however.

● ***Our adopted British Blue, Chips, grew up in a high-rise flat. We live in the country, but he hates going out. How can we encourage him?***

Give Chips a few weeks to get used to you and his new environment. If he looks out the window or shows signs of interest, carry him outside for a few minutes a day, and feed him in the doorway. In time he may start to explore outside on his own, but don't force him.

● ***We are ready to take home a cat from our local shelter, but the staff has warned us that she does not travel well. We are afraid she will be so traumatised that she'll run away as soon as she has a chance. Is there anything we can do?***

Try spraying her carrier with a pheromone spray (available from your vet) at least 10 minutes before you put her in it, and spray her new quarters daily while she settles in. Pheromone sprays help reduce stress in animals and have even been found to be useful for feral cats. Do not use the spray while the cat is in the vicinity as the sound or the freshly sprayed alcohol in the base may distress her.

● ***How can I tell the age of a rescued cat?***

You will have to take advice from the shelter staff – it can be difficult to judge a cat's age accurately. Experienced cat handlers and vets can tell if it is a kitten or less than a year old, or if it is very old. They can usually estimate if it is less than 5 years old. After that, the teeth, general condition and behaviour can be used as rough guides only.

◀ *A cat in a shelter waits patiently for someone to take it home. Not all such cats are strays – many have been given up by their owners because of domestic difficulties or illness.*

Choosing a Homeless Cat

- ✓ Is the cat healthy? Has it been vaccinated and neutered?
- ✓ Approximately how old is it? Are you looking for a kitten or young cat, or are you ready to take on an older cat?
- ✓ Is it a stray, or has it been living wild? If not, why did its owners give it up?
- ✓ Does it have any behaviour problems? These can take a long time to overcome.
- ✓ Is it housetrained?
- ✓ Is it friendly and outgoing, or does it show signs of timidity?
- ✓ Did it live indoors, or was it allowed out?
- ✓ Is it used to children or pets?
- ✓ If a longhair, does it tolerate grooming?

back to good health, do not be put off. But if the cat seems seriously ill, it is better to avoid choosing it, especially if you have other cats at home.

Infectious diseases such as feline flu, enteritis, feline leukaemia, and FIV (see pages 86–87) are found commonly in stray and feral cats. Flu and enteritis are commonly found in crowded shelters. Check what vaccinations the cat has had before taking it home. Not all shelters offer a full course of shots, so arrange for your vet to make up any deficit. Have the cat screened for FIV as there is no vaccine against this disease. Many shelters will insist that the cat is neutered before you take it home. If not, you should arrange to have this done by your own vet.

Settling Your Cat at Home

After bringing the cat home, it is best to confine it to the house for at least one week while it gets to know you and its new environment. If there are other cats or a dog at home, it's a wise precaution to keep the new cat in a portable crate while they are in the same room so that they have no opportunity to fight until they are familiar with each other's scents (see pages 48–49). Handle the cat frequently every day unless it is extremely timid, and talk to it and stroke it while it eats. Be patient. Many rescued cats have had bad experiences and will take longer than other cats to settle, but the reward of seeing them secure and happy is worth the time spent.

Choosing Your Kitten

▲ *There is nothing harder to resist than a kitten. Seeing it with its mother and litter will help you to choose the one that has the special quality that says 'I'm yours'.*

WHEN YOU HAVE DECIDED WHAT KIND OF CAT you want, it is time to go and choose your kitten. Many people automatically think of the pet shop as the source of their kitten, but in fact this is not the best place to look. Because of the crowded conditions, pet shops are ideal breeding grounds for disease and the operators may not be as careful in their choice of breeder as you would be. Also, in a pet shop, you cannot see the kitten with its mother in their home environment. This is a huge disadvantage in selecting any pet.

If your choice is for a pedigree kitten, you are much better off buying it directly from a breeder. This way you will get to meet the person who was responsible for rearing the kitten. You will be able see for yourself if the mother (sometimes the father, too) has a good temperament, and judge how well the kittens are being socialised.

If you have decided on a nonpedigree kitten, there are several sources you can go to. Check the noticeboard at your vet's, read the classified

What to Look For in a Kitten

- ✓ Does it have a lively, friendly personality?
- ✓ Is its mother (and father, if present) friendly, outgoing, and trusting with people?
- ✓ Is the coat glossy, with no sign of flea dirt?
- ✓ Are its eyes clear and bright, with no 'third eyelid' visible?
- ✓ Is its nose damp but free of discharge?
- ✓ Does it have white teeth, sweet breath and a healthy pink mouth?
- ✓ Are its ears clean, with no wax or discharge?
- ✓ Is the area under its tail clean?
- ✓ Does the abdomen feel slightly rounded but not hard, with no lumps?

ads in the newspaper, or ask friends and neighbours. You will probably discover several litters in your area to choose from – your biggest problem will be restraining yourself from taking home one or two from each. If you know of a particularly attractive, friendly, well-mannered cat in your neighbourhood that is shortly going to have kittens, it may be a good idea to wait a while and choose one from her litter. This is the best way of ensuring that your kitten enjoys a good start in life, with all the advantages of an even-tempered, outgoing mother. The owners may let you pick out your kitten when it is only a few weeks old and return for it when it is old enough to leave its mother.

Picking the Right Kitten

The most important thing about your kitten is that it must be healthy (see box). Many people feel sorry for animals that are ill, but you should not choose your pet out of pity. You may find you have to make frequent visits to the vet, and nursing a sick animal is time-consuming. Ideally, have the kitten examined by a vet before you agree to purchase it, especially if it is a pedigree. If the owner or breeder refuses to let you examine the kitten closely, you should look elsewhere.

The look of the kitten will obviously play a large part in determining your choice. But personality is just as, if not more, important. Unlike puppies, there is no such thing as a kitten that is 'too bold' – the bolder the better. If the kitten is old enough to move around (10–20 days) and has begun to be handled by the owners, it should be curious and not fearful. Do not swoop down on the kitten and pick it up too abruptly. Handle it gently, stroking it with your hands, and play with it a little before examining it for signs of ill health or lameness. A cringing kitten, or one that hisses or scratches, should be gently put back with the litter.

● ***When's the best time to separate a kitten from its mother?***

Not before age 6 weeks, but it is better to wait until 8 weeks, when weaning is complete. Taking away a kitten too early can contribute to behaviour problems (see pages 44–45, 66–67, and 82–85).

● ***Is it better to get a male or a female?***

Either a male or female can make an affectionate, rewarding pet (or a successful show champion). Cats of either sex that are not intended for breeding should be neutered before they reach sexual maturity (see pages 74–77). If not neutered, toms will wander, fight and spray, and females can be very vocal when in heat.

● ***Are there any advantages to having two kittens?***

If you are out at work for most of the day, one kitten on its own will be bored and lonely, and may grow up to be either destructive or aloof. Two kittens will keep each other company and be companions for life – but unless you give them lots of attention, they may be more closely attached to each other than to anyone in your family. The other important consideration is whether you can afford their necessary medical care.

● ***My children have fallen for the runt of the litter. Should I insist they choose a different kitten?***

That's up to you. Runts do have a greater chance of being sickly so why not ask your vet or an experienced breeder to check it over for you.

▶ *Handling a kitten gently is a good way to begin to get to know it. If it shrinks away from you, it may be timid. Also, watch it moving around or playing to make sure it shows no sign of lameness.*

Your New Kitten at Home

PREPARE FOR YOUR NEW KITTEN'S ARRIVAL BY getting all the necessary supplies and equipment in advance. When you bring it home, a familiar blanket and toy from its former home will help it feel less disoriented and insecure. Before letting the cat out of its carrier, ensure that all doors and windows are closed and any fireplaces or small recesses are blocked so that it can't escape. Put its food dishes and litter box in a quiet part of the house (but not too close together) where the kitten can use them without being tripped over by humans or intimidated by your other pets. Offer small meals at first and keep to its regular diet until it has completely settled in (see pages 16–17). Provide it with its own bed – a simple cardboard box lined with a blanket will do.

If you have waited until the kitten is 8 weeks old to bring it home, it should already be housebroken – you need only provide a litter box with the type of litter it is used to. Immobilise the pet door if you have one – the kitten should not be allowed outside until it has completed its full course of vaccinations (see pages 18–19).

Time To Settle Down

A good breeder will have made sure the kitten is already well socialised (see pages 44–45), but it's important that you introduce it carefully to its new environment, especially if yours is a busy family. If there are small children or other animals in the house, it may be a good idea to restrict it to one room at first. Allow time for the kitten to get to know everyone in a calm, relaxed atmosphere, and you are likely to have fewer problems later on. Even a confident kitten may be nervous when small, boisterous children are around, so supervise their encounters closely and do not let the children squeeze or handle the kitten roughly. Take it to the vet's for a pleasant social visit before you schedule its first checkup and shots. Short trips in the car at this age are also a good idea.

Give the kitten at least a few days to settle down before introducing it to other animals in the household (see pages 48–49). Do not expect instant friendship, though a kitten is more readily accepted by other cats, and more accepting of

▼ *Your new kitten is always ready for a game, and interactive play with a ball will help forge lasting bonds between you and your pet (see pages 52–53).*

Your Kitten's Basic Needs

- ✓ **A safe environment:** Remove all potential dangers, such as dangling electric cords and hanging curtains, from its path. Make sure cupboards, drawers, etc., are kept closed to avoid the kitten becoming trapped inside.
- ✓ **A calm atmosphere:** Provide a portable crate or indoor pen where it can be safe from other pets or noisy children but observe what's going on.
- ✓ **Its own space:** Put its bed, food and water bowls, and litter box in a quiet corner of the house or flat.
- ✓ **Continued socialisation:** Allow it to be carefully handled by other people and get it used to new noises and sensations such as washing machines, vacuum cleaners and so on.

them, than an adult cat in a new environment. The sense of smell is the most important key to familiarity and making friends. Try putting the new kitten inside a portable pen or crate in a corner of one room and let the other animals come and go in the room at will. That way you protect all parties and allow them to get used to each other gradually. At the same time, the kitten becomes used to all the normal domestic noises, which can be considerable in a busy family.

▼ *Your kitten is likely to need the following equipment: its own bed, a litter box and plastic scoop, fine combs and a soft brush, food and water bowls, some toys and a collar with identity tag.*

● ***Should I feed my kitten right away when we arrive home or should I let it settle down first?***

This depends on how well the kitten copes with the journey. If it is evidently stressed and has lost control of its bowels or bladder, do not feed it for several hours but let it have access to water in the meantime. However, if it seems reasonably calm and happy, let it settle for an hour or so, then offer a small meal of whatever diet it is used to eating.

● ***How much sleep does my kitten need?***

Plenty. Even adult cats spend up to 16 hours a day asleep, and kittens will catnap even more between their bouts of frenetic energy.

● ***Do we really need to provide a bed for our new kitten, as we are happy for it to sleep on our bed?***

It's best to train your kitten to sleep in its own bed. You may need to isolate it sometime – it may develop a contagious condition like ringworm or need nursing in a recovery crate after an injury. And what happens when you go on holiday? Your cat will be miserable if it is used to curling up with you at night. Make sure the bed will be large enough when your cat is fully grown.

Feeding Your Kitten

▲ *Your kitten will have the best possible start in life if you feed it a diet that meets its daily nutritional needs and ensures healthy muscle and bone development.*

Tips on Feeding

Do

- ✓ Use a complete balanced diet for kittens as recommended by your vet.
- ✓ Always keep a bowl of fresh drinking water available next to the food bowl.
- ✓ If you serve tinned food to your kitten, serve it at room temperature.
- ✓ Chop any table scraps into small pieces if giving them as an occasional treat.
- ✓ Dispose of uneaten tinned food after every meal (dried food can be left).
- ✓ Wash the food and water bowls and any serving utensils daily.
- ✓ Contact your vet if your kitten stops eating for 24 hours.

Don't

- ✗ Give your kitten raw meat or fish; always cook them thoroughly first.
- ✗ Give your kitten tinned foods intended for dogs or other pets.

IF YOU HAVE OBTAINED YOUR KITTEN FROM someone else's litter, it should be at least 8 weeks old and completely weaned when you bring it home. Most kittens are weaned onto commercial dry pellets or moist tinned food at about 3–4 weeks (see pages 82–85). Continue feeding the kitten its original diet when you first bring it home – a sudden change may be unsettling and lead to stomach upsets. The kitten will need four small meals a day until it is 3 months old, but by the time it is 6 months old, you should have gradually reduced the number of meals to two and be giving larger quantities. You can leave dry food out for the kitten to eat its fill as it wants, but if you feed it on a schedule, it is easier to tell if it is eating too much or too little. A kitten's calorie requirements rise steeply from 250–400 calories at 2–4 months to 425–500 calories at 4–5 months, levelling out at 600 calories between 5 and 6 months.

After a few weeks, if you wish to change the kitten's diet, introduce the new food gradually by mixing increasing proportions of it with the old food. There is no need to give your kitten milk – a proprietary diet formulated for kittens will provide the daily amount of calcium and other minerals required for healthy growth.

A Balanced Diet

Kittens have the same requirement as adult cats for animal protein, which is essential for healthy growth and development. However, some cat owners make the mistake of feeding their kittens exclusively on large quantities of red meat on the assumption that the more protein the better. This can be dangerous: an exclusive diet of lean red meat causes calcium and vitamin deficiencies, as does a fish-only diet. Too much liver can cause bowel problems.

The easiest way of feeding your kitten a balanced diet is to give it a complete commercially prepared food that will provide all the necessary nutrients. These come either as tinned or dry foods. You should always follow the instructions on the label, particularly if using a proprietary dried food. It is a good idea to give a diet especially formulated for the young or growing kitten, according to age. These are available from vets and most good pet shops. You may decide to complement the kitten's diet from time to time with a home-prepared meal, but do so sparingly as more than 25 per cent of its food given in this way will unbalance the kitten's diet. It may also encourage the kitten to become a fussy eater. Suitable treats are small portions of cooked fish, chicken, or egg, mixed with a little boiled rice or pasta. Never give your kitten raw meat or fish.

● ***My kitten, Puffball, is always hungry. Should I let her eat as much as she wants?***

No. Check the instructions on any packaged diet you are using and make sure you are feeding enough for the kitten's weight and age. Have you been neglecting the kitten's deworming treatment? Increased appetite is often a sign of worms in a kitten. If neither of these things is causing its constant hunger, ask your vet to examine the kitten.

● ***I'm out at work for most of the day. Can I use an automatic feeder to give my kitten, Max, his meals?***

Yes – provided you have checked that it works efficiently and that your kitten understands how to obtain its food. However, if you leave dry food out, the kitten can eat it when it wishes. Provided the food is changed daily, it does not have to be protected by an automatic feeder. Always make fresh water available.

● ***Are special kitten diets preferable to home-cooked foods?***

In most cases, yes. Because they are formulated for growth and provide the correct balance of nutrients, you can be sure your kitten's diet is not deficient in any way. Home-cooked or adult cat diets can be supplemented with additional nutrients and vitamins, but this is complicated and may lead to mistakes.

● ***Champagne, our Siamese kitten, loves to eat the dog food we give our Labrador. Can we feed it to her exclusively? Is it likely to harm her?***

I'm afraid so. Dog food has very low levels of taurine, which is an essential amino acid for cats – without it, they can develop eye or heart problems. It is a long-term risk – the problems can take 6–12 months to emerge while the cat uses up its existing reserves of taurine. Cats also have a much higher protein requirement than dogs, so her overall development may be affected. I advise keeping Champagne away from the dog's dinner if you can.

◀ *Kittens will happily lap up milk from a saucer, but though this was their traditional food, today's proprietary diets provide better nutrition. For drinking needs, kittens should be given water.*

Your Kitten's Health

Even if you have followed the guidelines for choosing a healthy kitten (see pages 12–13), you should arrange for a checkup within a day or two of bringing it home. Your vet may spot problems you were unaware of, and you should also discuss vaccination and deworming.

Newborn kittens are given natural immunity against disease from antibodies present in their mother's milk. This is so strong in the early weeks that there is no need to vaccinate them, but the kitten's natural protection begins to decline after the age of 7 weeks. By the time it is 9 weeks old, it should be ready for its first vaccinations, with a booster dose at 12 weeks. Shots can be given against cat flu, enteritis, feline leukaemia, rabies and chlamydia (see pages 86–87), which are all highly infectious. To avoid exposure to unvaccinated cats, keep your kitten indoors until the full vaccination programme has been completed. Other vaccinated family cats can be allowed contact with the kitten only if it has already been screened for feline leukaemia and feline immune deficiency virus.

Deworming should have begun at 4 weeks, when the kitten was started on solid foods. You must continue to deworm your kitten every 2 to 3 weeks until it is 3 months old, and then once a month until it is 6 months. Thereafter deworm your cat at least twice a year (if it is an outdoor cat) or have your vet do a stool test for parasites. Indoor cats should be checked once a year.

Routine Healthcare

- ✓ Make an appointment with the vet immediately for a general checkup.
- ✓ When it is 9 weeks and 12 weeks old, take the kitten for vaccinations.
- ✓ Give it a deworming pill every 2 weeks until 12 weeks old; then monthly until adult.
- ✓ When the kitten is 4–5 months old, ask your vet's advice on neutering. Most cats mature at about 6 months, so be prepared.
- ✓ Groom at least once a week, checking for fleas, and treat as necessary.

Getting Used to Grooming

An early start to grooming, essential for good health, gets the kitten used to being handled in very sensitive areas such as the paws, underparts and bottom. It also helps to build trust and affection. Check the kitten's ears and eyes, and gently wipe them clean with damp cotton wool. Comb the coat at least once a week with a small fine-toothed comb. Longhaired kittens need special attention under the legs, behind the ears and on the stomach, where knots and mats tend to build up. While grooming the coat, keep an eye out for flea dirt (see pages 88–89). If you find fleas, be sure

◀ *Don't forget to look in your kitten's mouth to be sure that its teeth, gums and breath are healthy. A second person can be useful for restraining the kitten if needed.*

● *When is the earliest that I should neuter my kitten?*

Puberty in most shorthair males and females is at about 6 months of age, and it is advisable to neuter or spay just before this. Longhairs mature a little later.

● *I bought a pedigree Siamese kitten from a breeder, and within a week of taking it home, I had to take it to the vet with a urinary infection. Should I ask the breeder for compensation?*

If your kitten becomes ill in the first two weeks after you brought it home, and your vet thinks it may have left the litter with the illness incubating, contact the breeder or original owner. If you paid a substantial amount of money, you have a right to some compensation.

● *When is it safe to let my kitten outdoors?*

Your kitten will not have full protection against infectious diseases for 7 to 10 days after receiving the second shots. When you do let it out, it is a good idea to go with it, carrying tidbits so that you can tempt it back with a food treat if it wanders too far.

to use an antiflea product recommended for kittens. Treat its bedding and other areas such as carpets to destroy flea eggs and larvae. Treat any other household cats at the same time.

Thinking Ahead to Adulthood

Before you know it, your kitten will be grown up. Females and Oriental breeds reach sexual maturity earlier than males and longhairs, some as early as 5 months of age. If you wish to breed your cat, you must be prepared to find suitable homes for the litter. Keep in mind that few people will pay for a nonpedigree kitten, and that in order for you to breed pedigrees, the stud and queen must both have registered status. Unless you plan to become involved in the business of breeding and showing cats, you should make an early decision to have your cat neutered.

▼ *Your kitten should be kept inside until its vaccination programme is ended. After that, you can consider the pros and cons of an outdoor life* *(see pages 30–31)**.*

House-training

CATS HAVE A WELL-DESERVED REPUTATION FOR cleanliness in both their grooming and toileting habits. Kittens are usually easier and quicker to house-train than puppies because they instinctively cover their faeces and urine with soil, and so are attracted to use the litter in their boxes. A mother cat usually teaches her kittens where to relieve themselves as she weans them onto solid food, when the kittens are between 3 and 4 weeks old. By the time your kitten is old enough for you to take home, you are unlikely to have to concern yourself with basic training.

The litter box should be kept well away from the kitten's food and water bowls, preferably in a quiet corner and easy accessible. Avoid putting the box in a busy hallway, near noisy domestic appliances, or close to where you prepare and eat food. Even if you intend to let your kitten have access outdoors so that it can relieve itself at will in a chosen part of the garden, you should still make sure that it is able to use a litter box indoors when necessary (see box). Then the cat will know what to do if it has to be kept inside following injury or illness.

Using the Litter Box

- Make sure you provide the same type of box and litter the kitten was using in its previous home to avoid unsettling it.
- Put the kitten in the box whenever it seems to feel the need, especially when it wakes up, and after meals. You may have to show it how to scratch the litter with its front paws.
- Reinforce good habits with praise or tidbits when the kitten performs in the box.
- Never punish a kitten by rubbing its nose in its mess – the smell will only reinforce the kitten's memory of the location as a toilet.
- Use a biological cleaner after 'accidents' in the house. This will stop the kitten from relieving itself again in the same spot.

Which Type of Box?

For reasons of hygiene, litter boxes should be made of a nonporous material that can be disinfected easily, such as metal, plastic, or fibreglass. Usually rectangular in shape, they should be about 36cm (14 in) long and (8–15cm) 3–6in high. Some commercial boxes are covered to conceal unwanted smells. If you use one, don't forget to change the litter daily. Disposable trays are available for convenience when travelling, at shows, or in catteries.

Various types of litter material can be used. Cats have individual preferences, but in general a highly absorbent commercial litter is best. Soil and sand can be messy, and sawdust is not very absorbent. Scented cat litters are more appealing to owners than to their cats, who may refuse to use the box if they dislike the scent of chemical 'fresheners'. Remove the soiled

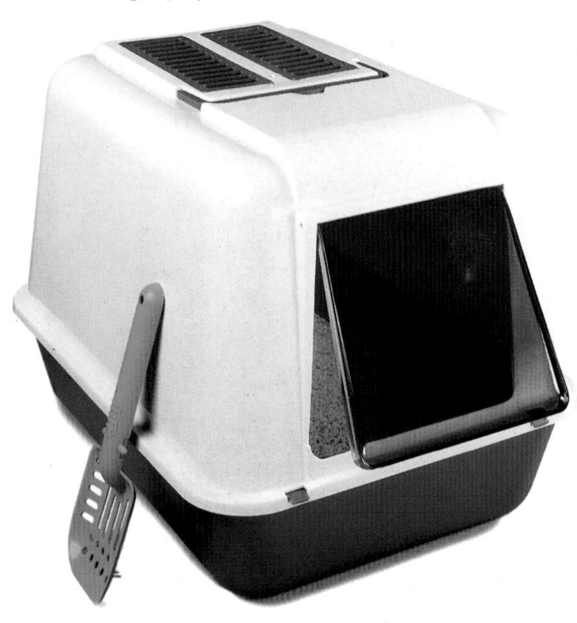

◀ *A cover on the litter box will help to minimise any unpleasant smells in the house, and your cat may appreciate the extra privacy. Remember to check and clean the box as regularly as if it had no cover; don't let it become a case of 'out of sight, out of mind'.*

litter as soon as possible, at least once daily and more often if two cats are sharing the box. Dispose of the faeces down the toilet. The litter box should be thoroughly cleaned and disinfected at least once a week. Pregnant women should never handle cat litter or boxes because there is a slight risk of toxoplasmosis affecting their unborn child (see pages 86–87).

Lapses in House-training

Always consult your vet if your house-trained kitten or cat suddenly begins to have 'accidents'. There may be any number of causes, including infection, chronic disease, or behavioural problems – for example, if the cat is very stressed for some reason (see Spraying and Soiling Indoors, pages 64–65). If your cat relieves itself inside the house, it is important that you clean the spot thoroughly to eliminate any odour. If any scent is left, the cat may be drawn back to use it as a regular toileting spot. Do not use an ammonia-based cleaner as this has the same smell as urine. Your veterinarian or a good pet store should be able to recommend a suitable biological cleaner.

▼ *House-training a kitten builds on the cat's natural desire to cover up its own waste. A 4-week-old kitten is taught by its mother to relieve itself on loose dirt or soil. The litter box continues this process.*

● ***I have recently adopted a lovely 3-year-old female cat, Suzy Q. She refuses to use the litter box, though she used one in the animal shelter where she lived.***

Are you using the same type of box and litter as she is used to? A change in her usual routine may be upsetting Suzy Q. Avoid scented litter if she is not accustomed to it. Some cats dislike being watched while they are using the litter box. Use a litter box with a lid, or put it under a countertop or table or inside a bigger box with one side cut away. Are other cats in the house using the same box? She may find this intimidating, so provide a separate one for her to use.

● ***Morris is behaving oddly around the litter box. He meows, scratches around, then urinates beside the box or on the edge. What could be wrong?***

Cats can stop using their litter boxes for all kinds of reasons – for example, if they were startled by a loud noise when using the box or if they have a sudden attack of diarrhoea. Have Morris checked by the vet for any infection or disease. He could have sore feet, which are being aggravated by large pellets or gravel in the box. Are you changing the litter enough, or is a sick cat sharing the box?

● ***My cat Xerxes has always used a litter box. Now he goes in other places around the house, passes small amounts of faeces, and only rarely uses the box. Why has his behaviour changed?***

You should take Xerxes to the veterinarian as soon as possible. Accidents of this sort could be caused by a disease of the digestive tract, possibly colitis, which prevents him from getting to the litter tray in time to relieve himself.

Feeding Your Adult Cat

From ages 6 months to 12 months, your cat needs 600 to 700 calories a day to maintain its growth energy levels; after that an average adult cat requires only about 450 calories a day. This is equivalent to about a large (500g/16oz) tin of cat food daily, split between two meals. Tinned food has the advantage of being available in a wide range of flavours, but many owners dislike the smell, especially in hot weather. Many complete dry diets are available for cats. These can be put out once a day for a cat to help itself, but the success of this depends on your cat's ability to regulate its own appetite. Most complete diets should be fed in smaller quantities than tinned food, but check the label on the package for instructions. These diets usually come in different formulas for growing, full-grown, overweight, or elderly cats.

Another advantage of dry food is its texture. Because cats use their teeth to tear rather than chew, most of them prefer food with a slightly lumpy texture. Dry foods help prevent tartar (brown scale) from building up on the teeth.

Cat breeds do not vary much in size, and the average adult should weigh between 3.5–5.5 kg (8–12lb). If your cat is being fed the right diet and is active, this should be easy to maintain. Weighing your cat regularly will help you notice if there is any significant loss or gain in its weight. Obesity can put a strain on the joints and heart. If your cat is very overweight – more than 7kg (15lb) – ask your vet's advice about putting it on a diet. Conversely, weight loss when the cat has a good appetite could indicate illness, and you should always consult your veterinarian.

Fussy Eaters

In general, cats are fussy about their food. They scavenge less than dogs and will walk away from anything that smells unpleasant by their standards. Also, as hunters, they have a marked preference for food that is at blood temperature. If you are giving your cat tinned food, or leftovers from your own dinner last night, do not serve it straight out of the refrigerator; heat it up first. The ideal temperature is about 35°C (100°F). Commercial cat foods are nutritionally balanced and are probably the best choice for your cat.

▼ *Cats have a reputation for being fussy eaters. You may inadvertently encourage finicky behaviour if you constantly switch diets.*

You can supplement a commercial diet with other foods such as tinned or cooked fresh meat, liver, fish, or cooked egg, but it is generally advisable not to give more than 25 per cent of your cat's food in this way. Once you form the habit of giving your cat home-cooked food, it may develop a preference for it, and it is unlikely to be as nutritionally sound as a specially formulated diet. Switching diets and making too many abrupt changes sometimes leads to diarrhoea. High-fat dairy products also tend to cause stomach upsets. Once a kitten has been weaned, it no longer needs milk. Be sure to provide your cat with plenty of fresh water, particularly when feeding it dried food.

Although cats are carnivorous, their natural diet in the wild is varied, including some vegetation – either in the form of grass eaten by the cat or the stomach contents of its herbivorous prey. However well fed it is, your cat will probably eat grass and, if it is able, catch whatever small prey it comes upon outdoors. It's also likely to show an avid interest in any food you are preparing or eating yourself. Most owners consider begging to be bad manners. If you wish to discourage it, do not give your cat tidbits from the table.

Home Cooking

Do

- ✓ Bake, grill, or boil meat, rather than fry it. Roast chicken is acceptable.
- ✓ Always remove any bones and chop the food into small pieces.
- ✓ Keep quantities modest and do not add too many vegetables or carbohydrates such as pasta.
- ✓ Offer sardines, tuna, herring, or mackerel for an occasional treat.
- ✓ Tempt an invalid or reluctant eater with cooked white fish.

Don't

- ✗ Add any flavouring. Many cats dislike spices, and a taste for sugar should not be encouraged.
- ✗ Give your cat raw meat. It may be diseased.

● My two nonpedigree cats, Nancy and Barbara, are large in frame. Now that they are age 12 and slowing down, their weight has crept up from 5.5kg (12lb) to 7kg (15lb). Should I put them on a diet?

Yes, you most certainly should. At 12 years old their joints will not easily be able to support the extra 25 per cent weight gain, and your cats will find it difficult to run and jump. They will become sedentary and even more obese. Arthritis may develop too. At their age the weight gain is also putting an extra strain on the heart and circulation. Ask your vet for advice about a weight-reducing programme.

● I've heard that you shouldn't use ordinary liquid detergent on animals' dishes. Is there any risk to their health?

It won't do them any harm, but cats have delicate palates and may turn up their noses if their dish smells strongly of detergent – always rinse thoroughly under running water to get rid of any lingering soap film or smell, and air the dishes well before using them again. You can always put the cat's dishes in the dishwasher, but be sure to run them separately from your own dishes for hygienic reasons.

1

▶ **1.** *Tinned food for cats will meet their daily nutritional needs but some owners (and cats) dislike the smell.* **2.** *A commercial dry food can be left standing all day for the cat to eat at will.* **3.** *Home-cooked food is best given as an occasional treat.*

2

3

Caring for Your Adult Cat

PUBERTY BEGINS AT ABOUT SIX MONTHS FOR your cat, and adulthood at 1–3 years (large or longhaired breeds may take longer; see pages 74–77). Although your cat is no longer a kitten, the daily routine of feeding, grooming and playing remains the same. Any changes you notice should be brought to your vet's attention.

One important difference between an adult cat and a kitten is that adults have the potential to breed – and they will breed if left to their own devices. Unless you have a pedigreed cat which you wish to mate, it is always advisable to ensure that your cat is neutered before it reaches sexual maturity (see pages 74–77). Cats that are not neutered, especially males, also tend to roam and fight more, and males will spray their territory, both indoors and outdoors, with pungent urine (see pages 56–57 and 64–65).

Cat Proofing Your House

You should try to ensure that your house is 'cat proof' against accidents. Cats are rarely clumsy but they explore wherever they can. Any fragile items should be put well out of a cat's reach, ideally in a cabinet rather than on an open shelf. Cats may crawl up chimneys, jump on a stove, bite through the extension cords on appliances, or knock over hot irons left on ironing boards.

● *Should we put our cat out at night or keep her in?*

Keeping your cat in at night is the best way to keep her safe, especially if you live on a busy street. If you live in a more rural setting and do not mind her going out, a pet door will let her come and go as she pleases, though you may receive 'trophies' after her nocturnal expeditions.

● *I have a 10-month-old Burmese, Baba, who loves climbing trees. I don't want to keep him in, but I'm worried he'll get stuck in a tree. What should I do?*

Don't worry. Most cats can get down from trees and will do so eventually if not harassed by a panicked owner. You can try to tempt Baba to come down with his favourite food. If he is up a tree longer than 24 hours, call your local fire department (not the emergency service!) and ask whether they can help.

● *Which plants are poisonous to cats?*

Some of the common ones are: several kinds of ivy, philodendrons, poinsettias, rhododendron, holly, clematis, sweet peas, bluebells, yew, lily of the valley, azaleas, delphinium, lupin and oleander. A complete list may be available from your vet. Remember also that chemicals used to treat plants may be eaten by the cat that eats the plants.

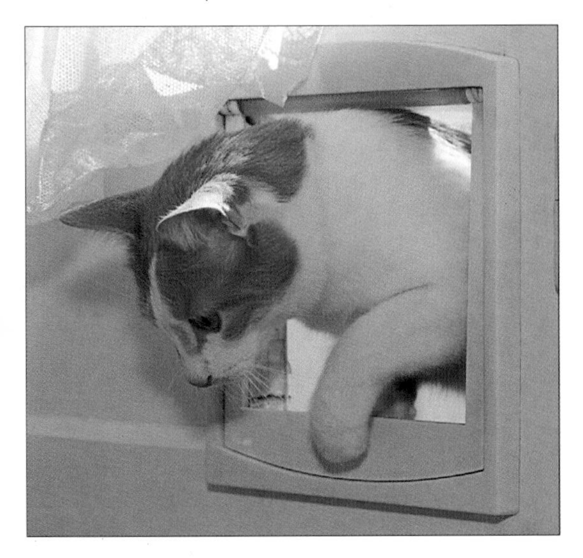

▲ *A pet door should be the right size and height for your cat to enter easily. Some doors lock from the inside or open when activated by a magnet on the cat's collar.*

◀ *Cats sleep more than most other mammals, as much as 16 hours every day, but not for long periods. This nap in the sun may last less than half an hour.*

Routine Care for Your Cat

- ✓ Feed your cat the correct diet (see pages 22–23). Always keep fresh water available.
- ✓ Weigh your cat every 6 months and make a note of it. Significant weight loss or gain is a cause for concern.
- ✓ Take your cat for booster shots every year (see pages 86–87).
- ✓ Check and treat for worms twice a year, or four times if your cat is an avid hunter.
- ✓ Groom at least weekly for a shorthair, daily for a longhair (see pages 26–29). Wipe eyes and ears gently with damp cotton wool.
- ✓ Check for flea dirt (see pages 88–89). Routine preventive flea control is advisable.
- ✓ Consider installing a cat-size pet door if you do not have one already.
- ✓ Minimise any potential hazards around your house and garden.

They may chew on a poisonous plant, crawl into a washing machine or dryer, or sit under a car – or, even worse, crawl beneath the bonnet from below and perch right next to the engine. Kittens and young cats are the most likely to get into trouble. Stop them going near any potential hazard with a sharp 'no!' and avoid tempting fate by closing doors, turning off appliances and putting things away. Do not leave windows open.

Cats, as well as kittens, require opportunities for games to prevent them from becoming bored and even destructive (see pages 52–53). If you are away from home for most of the day, it may be worth considering getting your cat a feline companion. A cat 'multigym' or activity centre provides exercise as well as entertainment and is particularly beneficial for an indoor cat.

If you do not wish to keep your cat indoors all the time, a pet door is very useful to allow it to come and go from the house as it pleases. If there is room, two doors are ideal: one leading into an enclosed porch or utility area and another with a lock leading into the house itself. A locked flap should stop unwanted 'gifts' such as mice and birds being brought into the house while you are out. It will also prevent unwanted visits from the more adventurous neighbourhood cats.

Grooming

MOST CATS GROOM THEIR COATS REGULARLY unless they are in pain, unwell, or very old. An unkempt appearance in a cat is often one of the first signs of ill health. Self-grooming keeps the coat clean, massages the skin, removes loose hair and stimulates hair growth.

Cats frequently swallow hairs they lick from their coats. These can build up in the stomach, causing fur balls. To prevent this, owners should regularly comb the loose hairs out of their pet's coat, particularly if it is shedding or losing hair due to flea infestation. Begin when the cat is a kitten to get it used to the routine. Start by brushing the cat gently with a soft brush when it is eating or is relaxed on your lap. Once it has learned to tolerate this procedure happily, you can start to use a comb. Remember to check the teeth, claws, eyes and ears regularly. Praise and stroke your cat all the time, and always finish each session with a tidbit or playtime.

If a cat has been neglected or refuses to be groomed, its coat may become badly matted, particularly if it is a longhair. Don't use scissors to cut away any tangles; you may injure the cat, especially if it struggles. Try using an antitangle conditioner. If this doesn't work, your vet may have to sedate it and clip the mats out for you.

While it is without all its fur, the cat will feel cold easily. Keep the cat indoors and out of draughts, or put a special manufactured coat on it. While the fur is still very short, begin gently brushing it with a baby's brush or a rubber glove. With a little incentive, the cat should learn to accept grooming.

Shorthaired Cats

Many owners find that grooming their cat helps relax them and enjoy doing so every day, regardless of whether their pet has a short or long coat. However, shorthaired cats in good health normally need grooming no more than once a week. This removes dead hair, prevents furballs, and means that your cat is less likely to leave fine hairs all over your carpets and furniture.

To groom a shorthaired cat, comb it from head to tail, then repeat with a bristle or rubber brush. If you stand your cat on a newspaper or white surface, it will be easy to see if any flea dirts (tiny black specks) drop off during this process. Finish the grooming by rubbing the coat with a chamois or other cloth to make it glossy. Use a soft baby's brush on the coats of Rex cats (see pages 154–155).

◀ *Begin every grooming session with a comb to ensure you reach all the way through the coat. Use an all-purpose fine-toothed comb for a shorthaired cat such as this one.*

▲ *A slicker brush is effective for smoothing out the hairs on the body of a longhaired cat. Start at the head and work your way down towards the tail.*

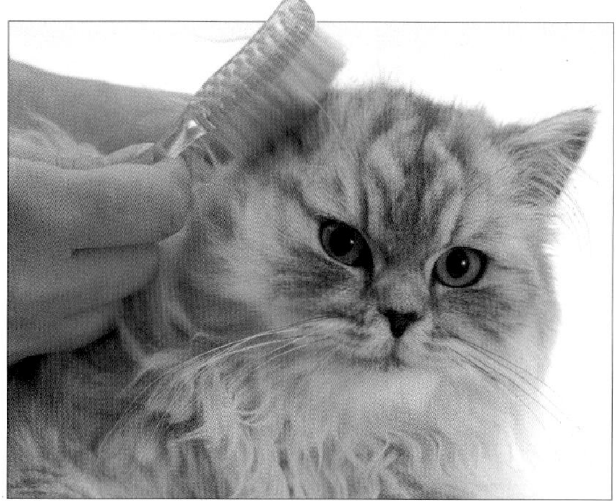

▲ *Use a baby's brush on the long hairs around the ears. These parts are difficult for the cat to reach and are particularly prone to tangle and mat.*

Longhaired Cats

The soft, fine hairs of many longhairs easily become tangled and form impossible mats. They require daily grooming to keep them free from knots. This is particularly true of Persians and their crosses. Other breeds such as the Maine Coon and the Balinese have glossy, flowing coats with very little undercoat. Not as soft and dense as the Persian's, they are much more manageable, and may need only weekly grooming.

Use a wide-toothed comb or slicker brush to tease gently through the hair all over the body. Take care not to hurt the cat by tugging too hard, particularly in very sensitive areas such as the abdomen and flanks. Check the underlying skin as you go along for signs of discoloration, injury, abnormal lumps and flea dirt. Tangles are found most commonly around the ears and behind the legs. Tease them out with your fingers and then comb through. Use a bristle brush for the tail, brushing gently to either side, and for the hair around the neck, if you like the 'ruffed up' look. If you have difficulty in combing through the coat, a spray-on conditioner should help get rid of tangles.

● ***Is it possible for a cat to groom itself too much?***

Yes. Cats can groom obsessively as a response to stress or if they have a medical or physical problem such as flea infestation or a skin condition (see pages 92–95). Take your cat to the vet if you see signs of excessive grooming, sore patches, or bald spots.

● ***My Chinchilla, Cybill, seems to get hairballs no matter how much I groom her. Is there anything else I can do to prevent them?***

Check with your vet that Cybill does not have a skin condition that is causing excessive hair loss. If all appears well, your vet may advise giving her a laxative such as a stool softener especially supplied for cats once or twice a week. Are you sure your comb and brush are the best ones for the job?

▶ *Your choice of grooming equipment includes wire and bristle brushes, baby's brush, guillotine nail clippers, slicker brush, wide- and fine-toothed combs and chamois cloth.*

● ***Why does my cat Daisy groom herself more in hot weather?***

A cat has to sweat through a thick coat, which means it doesn't lose much heat this way. By grooming itself, it dampens its fur with saliva, and this helps to keep it cooler.

● ***Florence is always scratching her ears, but I've checked and they look clean. What could be wrong?***

She may have a buildup of wax buried too deep to be seen. You can try to dislodge it by gently rubbing her ears and then following the steps for ear cleaning (see right). If the scratching continues, make an appointment with the vet to remove the blockage.

● ***I've heard that giving my cat a regular bath will prevent fleas. Is this true?***

No. The only thing that prevents fleas is regular treatment to kill any fleas present and prevent their eggs from hatching. Fleas and worms go together, so be sure your cat is dewormed (see pages 88–89).

● ***Shadow, my tomcat, loves to crawl through dirty places, and he hates water. How do I clean him?***

Unless Shadow is filthy down to the skin, which is unusual in cats, the best way is to wipe down his outer coat with a moist towel or baby wipes. This should be a less traumatic procedure for all concerned than attempting to bath him.

▲ *Clipping the nails is a sensitive procedure. Ask your vet or a professional groomer for advice if you are not sure where to cut. Take care to avoid the quick – it is better to take too little off than too much.*

Checking Problem Areas

If you groom your cat once a week, this is an ideal opportunity to carry out a full maintenance routine involving the ears, eyes, mouth and claws. If you groom your cat daily, you do not need to go through the entire procedure every time; use common sense. Remember that older cats may need more grooming as they become less flexible and able to reach for themselves.

1. Examine your cat's ears. Discoloured wax, inflammation, or an odour may be a sign of infection; normal ear wax should be honey-coloured and sparse, and should not smell offensive. Use an ear cleaning solution in the ear canal and clean away excess wax with cotton wool. Never poke cotton buds down a cat's ear canal.
2. Check the eyes for discharge or swelling and redness and consult your vet if you suspect any abnormalities. Bathe the lids with cotton wool dampened with a special eye cleanser for cats or with a saline solution.
3. Open the cat's mouth and check its teeth for plaque and the gums for any inflammation. Also look at the tongue and the roof of the mouth for injury or ulceration, then brush the cat's teeth. (Follow the instructions on pages 104–105 for brushing your cat's teeth.) If your cat has not become accustomed to this from an early age, and resists violently, you may both be much happier if you schedule a yearly session with the vet to have the cat's teeth descaled and polished. The cat will need to be anaesthetised for this.
4. Check the cat's paws for accumulated dirt. This is usually removed by the cat during its own grooming routine, but this may be more difficult in wet weather or if the cat is old and stiff. If the paws are caked with dirt, you can use damp cotton wool to clean them.
5. Active outdoor cats will keep their claws sharp by scratching trees and fenceposts. Indoor cats may need to have their nails clipped, as do older cats and those who tend to claw the furniture. Check regularly as overgrown claws can grow into the paw pad and become infected. Use guillotine clippers as these do not squeeze the nail bed, and get someone to help you restrain the cat. The nails are usually held in a retracted position, so in order to trim them they must be extended by pressing behind each claw in turn.

Cleaning a Contaminated Coat

If your cat's coat becomes accidentally contaminated, wrap it in a large bath towel or fix an Elizabethan collar to prevent it from licking the area and spreading the problem. You can attempt to clean the coat in the following ways, but if you are in any doubt, take the cat to your vet without delay.

PAINT, VARNISH Allow to harden, then cut away the hair carefully with scissors. Do not use paint solvent or thinner, as they can damage the skin.

OIL Cover the affected area in a vegetable cooking oil, or smear with a hand-cleaning jelly. Wash it off with warm, soapy water (liquid detergent will do) and rinse thoroughly.

TAR Cut away contaminated hair, or rub with vegetable cooking oil and treat as for oil (above).

Only the tips should be clipped, and the pink, sensitive 'quick' must be carefully avoided. If you are unsure about where or how much to clip, ask a professional groomer, vet, or veterinary nurse to show you the best technique.

Bathing

Unless your cat is a show cat, it should not need bathing under normal circumstances, but a bath can be very useful if your cat has crawled through something foul-smelling. No cat, except a Turkish Van, enjoys the procedure, but it can be done fairly easily if you are careful. Get someone to help the first few times, and if possible give the first bath when the kitten is only a few months old. Use warm water, testing it with your elbow – the ideal temperature is 38.6°C (101°F), the same as your cat's body temperature.

Either the kitchen sink or the bathtub is a good place. A spray attachment is very helpful. Use a rubber mat so the cat won't slip, and don't make the water too deep; 5–10cm (2–4in) is plenty. Gradually wet and shampoo the fur, massaging into the skin. Use a mild baby shampoo or special cat shampoo, but avoid all products labelled for use on dogs. Rinse until the water runs clear to prevent the cat licking off any shampoo. Dry the cat thoroughly with a towel and place in a warm room, or use a hair dryer on a low setting if your cat doesn't mind it. Keep the cat out of draughts until its fur is completely dry.

▶ *After a bath, wrap your cat in a large, warm towel and gently pat it dry until its fur stops dripping. Some cats may tolerate a hair dryer, while others consider it the final trauma.*

An Outdoor or Indoor Life?

▲ *If your cat is very active, outdoors may be the right place for it. An outdoor cat has freedom to roam and hunt but is exposed to more hazards.*

CATS ARE LITHE AND ATHLETIC ANIMALS WHO enjoy exercise. Many cat lovers believe that it is cruel to keep a cat restricted to the house all its life. However, other owners may consider that the risks of allowing their cats outside are unacceptably high, or they may live in high-rise flats with no access to the outdoors.

If all circumstances are equal, it is up to you to decide whether to keep your cat indoors or out. There are advantages and disadvantages to both lifestyles, but don't impose any sudden changes on your cat once it has adapted to one or the other. If you want an indoor cat, make this decision when you first bring your kitten home or find an adult whose habits fit in with your own.

Outdoor Cats

These cats can roam freely, run, climb, and chase birds and mice. Consequently they are less likely than indoor cats to become bored, frustrated, or obese. On the other hand, they are more at risk of street accidents (especially kittens and elderly or deaf cats), fight injuries and diseases transmitted by other cats. Pedigree cats may become the targets of thieves. It is highly distressing when a cat is missing and the owner does not know whether it has been stolen, taken in by another household, or even killed. Suburban and rural cats are at lesser risk, but can still incur injury, hypothermia, or heatstroke. It is advisable to have a 'pet door' that allows the cat to come and go as it pleases and seek shelter from storms, or during spells of cold or hot weather.

Indoor Cats

An indoor existence keeps a cat safe from all such hazards but raises potential problems of a different kind. All too easily, the cat lacks stimulation and activity, leading to behaviour problems such as aggression or furniture clawing. Unneutered toms may spray urine in the house, and unspayed females may urinate more frequently when in oestrus, in addition to becoming restless and howling all day.

You can make an indoor cat's environment more interesting by building an activity centre or indoor gymnasium out of strong cartons with holes cut in the sides, large cardboard tubes to run through, and a climbing tree or ropes to clamber up. A scratching post is a must if you want to spare your furniture (see pages 58–59). If you live in a high-rise flat, put screens in the window frames to stop the cat from crawling out through an open window.

Indoor cats tend to spend more time interacting with you through play and physical contact than cats that spend much of their time outdoors. However, keeping a cat indoors will not necessarily guarantee a high-quality relationship (see pages 46–47).

▶ *A small tray of grass will satisfy the indoor cat's need for vegetation. Lawn grass, catnip, parsley, sage and thyme are all suitable.*

Some Hazards in the Home

Your home contains many potential hazards for a curious kitten or cat. To avoid accidents:

- ✓ Disconnect all electrical appliances when not in use. Use safety covers on sockets.
- ✓ Keep the cat away from the stove. Cats can burn their paws jumping onto hotplates.
- ✓ Cats love to climb into dark places and hide. Keep the doors of washing machines and tumble dryers closed when not in use. Check drawers before closing them.
- ✓ Make sure sewing accessories aren't left lying around. Wool and threaded needles are easily swallowed.
- ✓ Keep all household cleaners out of reach.

● ***Minky keeps attacking our feet and legs when we move around. He is an indoor cat who sleeps all day while we are out. What is this fetish about?***

After sleeping all day, Minky is ready for action when you come home, and your feet have become his prey. Provide some gymnastic equipment and offer him some stimulating games before and after work. Use toys that roll or can be dragged away from your body so that he pounces on them, not on you.

● ***My cat, Miou-miou, is 8 and has always lived indoors. Now that we have moved to the country, is it safe to let her out?***

Yes, but she will probably be very timid and unadventurous until she becomes used to the sights and sounds of the world outdoors. Take care she doesn't panic and run away. Go with her at first, and tempt her back with a tidbit. Some cats choose never to venture beyond the house or garden – don't force her if she would rather stay at home.

Transporting Your Cat

THERE CAN BE FEW MORE UPSETTING EVENTS for an independent, territorial-minded cat than being seized, confined and forcibly taken to an unfamiliar place – whether on a visit to the vet or to a new house in a new neighbourhood. Whenever you are taking your cat anywhere, it is important to minimise the stress as much as possible.

Your cat's safety must be your first consideration when planning how to move it from one location to another. A frightened animal will often behave unpredictably and can run away or lash out. For this reason, cats should always be transported in a carrier. Unless designed for carrying a small animal, cardboard boxes are not escape-proof, and may become wet and unsafe if exposed to rain or if the cat urinates. Wicker and mesh baskets look attractive but may require draught-proofing in cold weather; wrap newspaper or polythene around the outside to insulate the basket.

The most practical option is a fibreglass or plastic carrier. It will last for years, is secure and strong, and can be easily cleaned and disinfected.

What To Put in the Carrier

For the short journey a blanket and toy will suffice, but for the longer trip you may need the following:

- ✓ A familiar blanket: heavy for winter, lightweight for summer.
- ✓ The cat's favourite toy for entertainment.
- ✓ Dry food – either in a bowl inside the carrier or offered by you during rest stops.
- ✓ Water bowl, refilled at intervals.
- ✓ Waterproof lining on the floor in case of spillages or if your cat urinates.
- ✓ Torn up newspaper strips or litter tray in case the cat needs to relieve itself.

Another option, ideal for carrying a cat on public transport, is a large vinyl zipper bag with ventilation holes and a 'window' at one end. Some cats seem to feel more secure in the dark, so a blanket placed over an open mesh basket may make your cat less afraid. Others appear to enjoy being able to see their surroundings.

When travelling with your cat in the car, do not be tempted to let it out of its carrier. A cat on the loose in a car presents a major distraction to the driver. The carrier must be properly secured, either by a seat belt or by placing it on the floor so it will not slide around. On long journeys, the noise and motion of the car usually have a calming effect, and most cats settle down to sleep. However, if your cat is a difficult traveller, your vet may prescribe sedatives to help calm it.

For air travel, cats must have a special carrier that is approved by the airline. Call the airline well in advance to find out what is required, and make the necessary arrangements. If the flight is long, it may be worth having your cat sedated.

▲ *You can make your cat familiar with its carrier by making it part of everyday life around the house. Leave the carrier out in the open, door ajar, and wait for the cat to investigate. Put your cat's blanket inside it with toys or some food to encourage exploration.*

● *My two cats, Flotsam and Jetsam, are only small and don't take up much room. Can I put them in the same carrier to save space when travelling?*

So long as your cats are good friends and the carrier is sufficiently roomy to allow them to turn around, you can by all means use a single carrier on car journeys. However, airlines are likely to require the cats to be put in separate crates. Make enquiries first to find out what is needed.

● *Holly, our Burmese cross, is a bad traveller. We're planning to take her on holiday with us, but as we have a a long journey in the car to get there, we think she will need sedation. Does the vet need to do this? How long do the drugs take to wear off?*

Your vet will probably want to examine Holly before prescribing any sedative for her, but once that is done, you can give it to her yourself in pill form. The sedative should take about half an hour to work and its effects last for up to 8 hours.

● *Should I feed my cat before a trip? If so, when?*

If your cat is fed just before setting out it may suffer from motion sickness. It's best to wait at least one and a half hours after a meal before starting a journey.

Moving to a New House

Cats typically adapt quickly to new situations and most will normally accept a temporary or permanent move to a new house. Keep your cat indoors for the first few days to give it time to settle. A cat placed in strange surroundings may run off to look for familiar landmarks and be unable to make its way home. Wait until it is hungry, therefore, before letting it outside. If it seems inclined to wander, you can then call it back to the house with an offer of food. Make sure that the cat's identification disc has the new address and telephone number on it, even if you have taken only a holiday let. Your cat is likely to take a while to establish a position in the hierarchy of neighbourhood cats, so if your stay is going to be a short one, you may prefer to keep the cat inside to avoid squabbles.

▼ *A zipper bag is suitable for carrying one cat. Smaller and softer than most carriers, it has a shoulder strap to leave your hands free, but it may not be the best choice if your cat tends to be a nervous traveller.*

Holiday Care

IF YOU GO AWAY ON HOLIDAY, WHO WILL TAKE care of your cat? If your trip is only overnight or just for the weekend, your cat may be happy to be left to its own devices until your return. Most cats do not like change in their lives, and staying at home even without their owner is much less stressful than being moved into a boarding kennel. Electronic cat feeders, which measure out the right amount of food at scheduled intervals, may be used to make sure your pet is fed regularly. Be sure to leave out plenty of water as well.

What happens, however, if your cat gets sick or has an accident? If no one discovers the problem until your return, by then it may be too late. For a trip that lasts more than a day or two, it's best to arrange for a member of your family to stay behind or move in for the duration, or for a neighbour to come and check on your cat. Another option is to hire a 'pet sitter' to stay in your house or visit daily. This can be expensive, but it has the additional benefit of providing security for your house. The best recommendations come from people you know. Enquire at your vet's or among your friends. Always check references and meet the pet sitter in advance to establish that he or she appears trustworthy.

● *How long is too long to leave my cat by itself?*

This depends on what arrangements you have made for someone to come and feed it. Most cats can certainly be left overnight with an automatic feeder and an adequate supply of water. One week is too long to leave your cat without someone coming in at least once a day to check on it, clean the litter box, fill the water bowl and add food to the feeder. If your cat is likely to be miserable, this is another factor to consider; it may need scheduled play and petting sessions.

● *When visiting my local cattery prior to going away, I was told that all the cats were fed dried food. William usually has tinned food, and I am worried this may upset him. Should I put him somewhere else?*

You are right to be concerned. A sudden change of diet may upset a cat's digestion, causing diarrhoea and possibly vomiting. I would be very suspicious of any cattery that wants to dictate to you what your cat is to be fed, and I would advise you to look elsewhere for somewhere to put William while you are away.

● *My cat, Felicia, is going into a cattery while I am abroad for a month. I'm concerned that I have not been asked to have her vaccinated against feline leukaemia. Is she at risk?*

Your cat is unlikely to catch leukaemia in a kennel where she has no direct contact with other cats. The virus that causes it is usually spread through bites, mutual grooming and sharing food bowls with infected animals. Of course, you can have Felicia vaccinated anyway.

◀ *An automatic feeder may consist simply of a closed bowl that opens when the cat steps on the lid. Or it may be a device with a timer.*

▲ *A cattery will care for your cat if you are away for a long period. Always make arrangements in advance: good catteries tend to fill up quickly. All your cat's vaccinations must be up to date or most reputable catteries will refuse to accept it for boarding.*

Which Cattery?

When going away for more than a few days, you may decide to put your cat into a cattery. Don't feel you have to use the first one you visit if you don't like what you see – there should be plenty of choice, wherever you live. Personal recommendation is an excellent way to find out which catteries are the best. Alternatively, your vet may have listings. Ask to look over the premises and make sure you are satisfied with all the facilities before deciding to place your cat there.

Before leaving your cat, check that its vaccinations are up to date. It is also a good idea to give a last-minute preventive flea treatment, as fleas can spread rapidly where there are other cats in close proximity. Leave your vet's name and, if possible, a telephone number where you can be reached in case of emergency. To help make your cat's stay less stressful, inform the cattery staff of its normal diet and tell them about any particular likes or phobias it may have. Always inform the staff beforehand if the cat has any medical problems and, if so, what the symptoms are.

What To Look For in a Cattery

- ✓ Clean, light, well-ventilated facilities.
- ✓ Pens large enough to provide exercise and stimulation.
- ✓ Individual sleeping areas, quiet and warm, where your cat can retire for security.
- ✓ No cats admitted without proof of vaccination for cat flu, feline enteritis and rabies (where relevant).
- ✓ No direct contact allowed between your cat and anyone else's.
- ✓ Isolation units provided for animals that become ill while on the premises.
- ✓ Alert, helpful staff who are knowledgeable and appear to enjoy their work.

The Older Cat

In recent years cats have become ever more popular as pets. At the same time, the standard of feline health care has improved, leading to a greater population of older cats. A cat is considered elderly after about 10 years of age, but an average lifespan is 13–14 years, with a few cats surviving into their late teens or twenties. With most cats the decline into old age is gradual, and few major changes to their routine are required as long as the cat remains healthy.

All cats should have annual vaccinations and receive regular checkups so that any problems are identified and treated at an early stage. A cat that is more than 10 years old should be thoroughly examined by your vet twice a year. With advanced age the risk of developing disease is increased. It is a common mistake to attribute changes in your pet to aging when they may be due to an underlying medical condition that can be treated. Because older cats are more likely to have more than one health problem, your vet may advise a blood test before giving any major treatment to ensure that there is no additional condition that may complicate the treatment. A more detailed look at your cat's health will ensure that treatment for one condition will not worsen another underlying problem.

Problems of the Older Cat

Elderly cats often become thinner. Weight loss can be a sign of a medical problem such as an overactive thyroid gland. It can also be caused by a decline of kidney or liver function. Though some cats can lose more than half their kidney function without apparent effect, a failing liver is a more serious condition. Consult your veterinarian, but if no medical cause can be found for your cat's weight loss, it may simply be eating less because its sense of smell and taste has become less sensitive, leading to a poor appetite. You may be able to encourage a reluctant cat to eat by warming its food to blood temperature (about 35°C, 100°F). Feeding small, frequent meals with plenty of variety is better than giving

◀ *Old age can begin at around 10 years in a cat but often starts later than this. As cats age, they tend to become more sedentary in their habits and spend a lot of time during the day asleep or simply taking it easy. A decline in activity is frequently the first indication that your cat is growing old; cats do not show many obvious physical signs of aging.*

one or two large meals a day. The ideal nutrient supply for the older cat is the subject of continuing research. What is known is that the gut of an older cat is less efficient in absorbing nutrients, and veterinary diet experts now believe that reducing the levels of protein and phosphorus makes digestion easier. Your cat may benefit from being put on a commercial diet specially formulated for older cats' digestion, with less protein and more vitamins.

Constipation (see pages 106–109) is another typical problem in elderly cats. It can be prevented by adding fibre or petroleum jelly to the diet. Treat acute cases with a dose of liquid paraffin. An older cat may find it harder to control its urine and bowel movements, or be too stiff or tired to go outside. If yours is an outdoor pet, provide an indoor litter box in an easily accessible area, especially in winter.

You should check your cat's mouth regularly for infected teeth and gums (see pages 104–105). A sore mouth or stiff limbs sometimes interfere with grooming, so you may need to lend a helping hand. Wiping its coat gently with a damp cloth and combing it with a fine-toothed comb will help your cat look and feel better.

Needs of the Older Cat

- ✓ **Vaccinations** should still be given yearly.
- ✓ **Regular checkups** will help the vet spot if there are any new problems developing.
- ✓ **Diet** should be specially formulated for the older cat.
- ✓ **Add a teaspoon of bran** to the food once a day to prevent constipation.
- ✓ **Grooming** should be more gentle and more frequent if your cat seems to be having any trouble keeping itself clean.
- ✓ **Mouth** should be checked regularly for signs of soreness or infection.
- ✓ **Bed** should be put in a warm place, away from any draughts in the house. Padding will keep stiff joints more comfortable.
- ✓ **Litter tray** should be provided, especially in cold weather, in an easily accessible place.

▲ *An older cat with stiff joints will appreciate getting as close as possible to a heat source. A cat bed designed to be slung over a radiator makes a real treat. The soft fabric should be washable.*

● ***Oliver, my 20-year-old Siamese, has taken to sitting in the middle of our street, oblivious of traffic. Why do you think he does this?***

The simplest and most probable explanation is that Oliver has become a little deaf, blind and senile. In view of his exceptional age, this isn't really surprising. It's a shame to restrict him, especially if Oliver loves going outside, but I don't think you have any choice.

● ***Wizard is 14 years old. Though he has developed a ravenous appetite, he seems to be losing weight, and once or twice he has had an accident in the house. What could be causing these symptoms?***

Wizard may have an overactive thyroid gland, quite common in older cats. The thyroid gland controls the cat's metabolism. If it is overactive, the metabolism speeds up, making the cat hyperactive and restless. The cat will lose weight despite eating more than normal, and diarrhoea is also quite common. Take Wizard to your vet – a simple blood test will establish the levels of thyroid hormone in the bloodstream. Removing the thyroid gland should solve the problem.

● ***In the past Jessie, my 11-year-old cat, never had problems with her claws, but now they are frequently overgrown. Why is this?***

Jessie may be getting less exercise and grooming herself less frequently, but hyperthyroidism sometimes causes nail growth so have your vet examine her. Either clip Jessie's nails yourself or take her to the veterinarian for regular clipping.

Death of a Pet

THE AVERAGE LIFE SPAN OF A CAT IS 15 YEARS. If you own a cat, chances are you will have to come to terms with its loss someday. Most cats do not die of old age but need a helping hand from the vet. Euthanasia is the medical term for 'putting to sleep'. It offers a peaceful end for animals who are terminally ill, incurably in pain, or have lost essential body functions and cannot live an active life. All too often owners avoid taking an old cat for treatment until it is too late, for fear that the vet will suggest euthanasia. No vet likes to lose a patient, and where treatment is available it will be given. A vet may euthanise an animal only with the owner's consent. Conversely, the owner may disregard the vet's recommendation and request euthanasia for the cat. If the vet disagrees, he or she may refuse and refer the owner and patient to another vet. No vet may take a cat away from an owner and find it a new home without prior consent.

What Happens in Euthanasia

In almost all cases, the cat's transition to unconsciousness takes place smoothly almost as soon as an overdose of anaesthetic is injected into a vein in the animal's right foreleg. The cat is held gently and may be sedated first, if nervous. The familiar presence of the owner will help to reassure and calm it, but some owners cannot bear to watch the procedure for fear of the distress it may cause. This is natural, and the vet will sympathise with your decision. However, being present often helps to alleviate your grief as you can see for yourself just how peaceful the cat's final moments were. The cat is usually unaware of the injection, and the effects are so rapid that it falls asleep in five to ten seconds. Breathing ceases within a minute or so, and shortly after this the heart stops. There can be no kinder way of ending life and suffering. Very occasionally, the cat's blood pressure will be so low that the injection is given into the abdomen. This is equally painless, but the cat takes longer to drift off to sleep.

Your vet will be able to advise you about disposal of the body, and may offer facilities for cremation and burial. Many owners prefer to inter their pets on their own land. The body should be placed in a cardboard or wooden box and covered with a layer of soil at least 0.6m (2ft) deep.

Coming to Terms with Loss

Grieving for the loss of your pet is completely normal and natural; you will feel better if you do. You'll find that friends who have pets themselves will readily understand your feelings. Also talk to the vet or nurse a few days later if you think this will help. Don't blame yourself unnecessarily for your cat's passing. If the decision was carefully considered and made with your vet's advice, euthanasia was the correct choice.

Assessing Quality of Life

To decide whether euthanasia is necessary or not, your vet, in discussion with you, will assess whether the basic needs of the cat are being fulfilled. A life without suffering requires:

- ✓ Freedom from any form of pain, distress, or discomfort that cannot be controlled by medication.
- ✓ The ability to walk and balance freely.
- ✓ The ability to eat and drink without pain or vomiting.
- ✓ Freedom from painful, inoperable tumours.
- ✓ The ability to breathe without difficulty.
- ✓ The ability to urinate and defecate without pain or difficulty.
- ✓ The ability to see or hear well enough to cope with daily living.
- ✓ An owner who is able to cope physically and mentally with any nursing that may be needed.

If your cat lacks any of these essentials, and if examination and discussion with your vet reveal that treatment is not possible, then it cannot live a normal happy life.

● *My elderly Scottish Fold, Clyde, is losing weight and spends most of the day sleeping next to the boiler. I am hoping he will die in his sleep. What should I do?*

Take Clyde to the vet for a checkup. He may have failing kidneys, diabetes, cancer, or thyroid trouble. The vet may be able to prescribe treatment that could improve his life for a few more years. However, if the illness is serious and terminal, you should consider euthanasia. A natural death can be very unpleasant.

● *Should I tell my young children that I had to put Puffball, our ancient tabby, to sleep?*

Yes. You should always be honest with children about the death of any pet. Explain the reasons why Puffball had to be put to sleep and show them that it is right and natural to be very sad for a while.

● *My beloved 13-year-old Angora, Mimi, died three months ago, and the house seems so quiet without her. Is it fair to get another cat so soon?*

It is perfectly normal for you to be grieving still, but some people find that they want to get another pet sooner rather than later. There is nothing unfair about this, unless you start comparing the old one and the new one. Nothing can ever replace a lost pet, but a new one will have its own charms. You might want to get a kitten, which will demand more of your time.

▶ *A treasured photograph helps to keep alive memories of a much-loved friend and companion.*

▼ *Some owners choose to memorialise their cat with a headstone in a pet cemetery or private burial plot surrounded by other family pets.*

YOUR CAT'S BEHAVIOUR

CATS HAVE A WIDESPREAD REPUTATION for doing exactly as they please. Some people admire this independence and cite it as proof of feline high intelligence; others deplore it as proof of cats' innate selfishness. Their reputation for independence arises from their habit of hunting alone rather than in packs. However, although cats are not usually thought of as social animals, in the wild they often live in cooperative groups or colonies – in the same way as a pride of lions does. Domestic cats are certainly capable of developing close bonds with humans and even with other household animals, as any cat owner can attest.

In order to live in harmony with your cat, it is well worth making the effort to understand its natural instincts. If you can recognise its normal patterns of behaviour and know what causes them, you will be better able to find the right balance between educating your cat to fit your lifestyle and allowing it to follow its instinctive nature. It is easier to persuade a young kitten to adjust to your way of doing things than an adult cat, but the latter can be done if you patiently encourage your pet with repetition and reward.

You will stand a greater chance of achieving a balanced relationship if you work with your cat's nature rather than against it. Time spent in preventing behaviour problems now will be amply repaid later, enabling you to share your house and life with a content, well-mannered and endlessly fascinating creature.

How Cats Communicate

CATS WERE DOMESTICATED LATER THAN DOGS, and their transition from wild hunters to sociable companions is not complete. They still communicate according to the rules and signals that evolved in their colonies in the wild. Respect for one another's space and privacy is the basic principle that allows cats to coexist with others. They communicate with each other using signals that range from the very blunt to the very subtle and employ senses much more acute than those possessed by humans.

Smell, touch and hearing are probably the most important. They enable the blind, newborn kitten to identify the characteristic scent of its mother and grope its way over to her for nursing. The vibrations of her purring tell the kitten

- ***After my cat Priscilla was attacked by a dog, she smelled disgusting. Was it her or the dog?***

Scent sacs under the tail contain pungent-smelling matter that is discharged if the cat becomes very stressed or frightened. This is probably what happened when Priscilla was attacked.

- ***What do cats use their whiskers for?***

The cat's sense of touch is concentrated in the nose, feet and sensory hairs, which are thicker and stiffer than normal hairs. The most obvious of these are the whiskers, and they are extremely important in conveying information to the cat about its environment through touch and stimulation.

- ***Is my cat trying to tell me something when it turns its back on me after I scold it?***

Yes. Your cat is acknowledging its inferior social status and is avoiding an escalation of conflict by cutting off eye contact between you. It is not showing guilt.

▼ *A kitten learns the vocabulary of feline communication through its early contact with its mother.*

▶ *You can learn to read your cat's moods by studying its facial expressions. A cat that stares fixedly at you may be showing aggression. In normal daylight, dilated pupils indicate that the cat is agitated.*

that she is close by, and if it strays too far, its cries will be answered by the mother's chirp to call it back.

Touch and smell play an essential role in the way adult cats form and maintain relationships. Two compatible cats will often spend hours sleeping next to and grooming each other, and they typically greet each other by rubbing faces. As your cat is unable to reach up to your face most of the time, it greets you by rubbing itself against your legs instead, marking you with its scent and picking yours up. Then it walks away and begins licking its fur to remove your scent and restore its own. All cats continuously mark their territory with their individual scent, which is left by their paws, tails and faces wherever they walk, claw, or rub against something.

Expressions and Body Language

Visual signals are also important, both on their own and when combined with other forms of communication. A cat's face is easily read by another cat. Dilated pupils indicate arousal, usually fear or aggression, but sometimes they can indicate excitement. A fixed stare is often the prelude to a fight.

The facial expression is amplified by the cat's body language, especially the position of its ears. These are folded back in a show of defence, flattened to show extreme fear, pricked up or held slightly forwards to show confidence. A threatened cat tries to make itself look larger by fluffing out the fur all over its body and arching its back. The tail may be curled around the body for protection. Crouching with an arched back is a preparation for attack, but with a flat back it shows apprehension and submission. A cat that rolls on its back and exposes its vulnerable underparts to you is giving you the ultimate proof of its trust.

Vocal signals are varied and usually obvious, ranging from hissing and growling to purring and many different kinds of meow. Most people believe purring indicates feline pleasure. While it's safe to assume that a cat purring on your lap is content, cats do purr in other circumstances. Some experts believe that the activity is used by cats to calm themselves when they are frightened or in severe pain, and in other encounters to demonstrate nonhostile intentions. The best way to get a cat to purr is to let it approach you in its own time and then stroke it, rather than abruptly swooping down on it to pick it up.

▶ *A frightened cat is easy to spot: it arches its back, and its hair stands on end in the classic 'scalded cat' posture. The position of the ears suggests that the cat is not confident about attacking and may turn and flee.*

Preventing Behaviour Problems

A BEHAVIOUR PROBLEM IS ANYTHING THAT THE owner (or other people in contact with the cat) considers unacceptable. Some common 'behaviour problems', such as spraying or fighting, are completely natural activities from a cat's point of view. They are a problem for the owner because they are a nuisance or cause embarrassment.

Most cat owners want to discourage behaviour that endangers the cat or any other member of the household, or damages their property. Even if the cat's behaviour is only an inconvenience, you may find yourself becoming annoyed with your pet, and you will both be happier if the problem can be resolved. Being firm with your cat does not mean being harsh. Punishing it by shouting, locking it out of the house, or hitting it will only upset it and exacerbate the problem.

Any well-behaved cat should be able to use a litter box or go outside to relieve itself, and tolerate grooming. Common but minor problems like nibbling houseplants or sneaking food from countertops can be considered a matter of teaching your cat good manners. In addition, there are a number of bad habits and forms of nuisance behaviour that owners wish to prevent their pet from developing, such as spraying and soiling indoors, chewing fabric, clawing furniture and aggression towards people or other pets. Ways of dealing with these problems are discussed individually later in this section.

Sometimes a persistent behaviour problem such as excessive self-grooming, self-mutilation, or inappropriate spraying may be a symptom of a physical disease. If your cat develops habits of this kind, you must first take it to the vet for a checkup. Once all medical problems have been ruled out, the vet may suspect that stress or a compulsive disorder is causing the cat's behaviour (see pages 66–67) and refer you to a behaviour specialist for treatment.

Encouraging Good Behaviour

Do

- ✓ Set realistic expectations for your cat's behaviour, especially if it has been adopted.
- ✓ Begin training your cat as soon as it has settled into your house.
- ✓ Be consistent in applying the house rules.
- ✓ Reward and praise good behaviour.
- ✓ Say 'No!' or squirt the cat with a water pistol if it misbehaves.

Don't

- ✗ Shout at the cat or lock it out of the house.
- ✗ Let the cat see you aiming a water pistol; it will begin to avoid you.
- ✗ Ever use physical punishment, which will distress the cat and may even injure it.

The Importance of Socialisation

Not all behaviour problems can be prevented. However, it is sensible for breeders to select only cats with good temperaments for breeding. Even more important is the care of the newborn kitten (see Raising Kittens, pages 82–85). The first two to seven weeks of life are the most crucial, because this is when kittens acquire their fundamental socialisation. What happens at this age will determine whether a kitten grows up friendly and relaxed in a variety of situations.

To minimise the possibility of problems developing later, young kittens in the litter should be handled several times a day for 5–10 minutes by several different people of all ages. The handling should include lifting the kittens off the ground and gently restraining them. After you bring a new kitten home the socialisation process continues while the kitten becomes familiar with its new environment.

It helps to make sure that the kitten or cat you choose is well-socialised and appropriate for your circumstances: a timid cat introduced into a household with three small children and a dog is likely to have difficulty adjusting.

Educating Your Cat

If the cat was correctly handled from an early age, basic manners can easily be taught through reward and repetition. A tidbit is a good way to encourage your cat to sit still for grooming, for example. To discourage any undesirable behaviour, interrupt the cat while it is in the act and say 'No!' If necessary, squirt it with a water pistol or a clean spray bottle filled with water. Try not to let the cat see you or it will learn that it only gets squirted when you are present.

Nervous or aggressive cats are harder to train, but a lot depends on your relationship. Training can improve it, since it means paying plenty of attention to the cat. This is very helpful if your cat is shy or aloof, or needs a lot of stimulation.

● ***How do I teach my cat Sidney to come when he is called?***

Appeal to his greed by feeding him several small meals a day of his favourite foods. Make a big fuss about filling his bowl and putting it down, and he should come running. Call out, 'Sidney, come', and soon he will associate the command with dinner. Practise calling him to you for a tidbit. In time you can substitute praise for food.

● ***My cat Bosie used to walk all over the kitchen worktops and the table. Now he only does it when I'm out – he leaves pawprints. Is there anything I can put on the surfaces that will keep him off?***

First of all, don't leave anything out, such as food or toys, that might tempt Bosie to jump up. Try covering the surfaces in aluminium foil. Cats hate the feeling of this slippery, crinkly material under their paws. Do this for several weeks, and then do it at random just to reinforce the lesson. This should solve the problem.

● ***Loli has been with us for 5 years, since she was 3 years old. She is perfectly well-behaved and friendly until we try to pick her up; then she always growls and even scratches. It's a nuisance when we have to groom her, put her in her carrier, or have her examined by the vet. How can we get her to stop doing this?***

If Loli has always been this way and your vet is sure there is no underlying physical condition such as arthritis, her behaviour is probably due to limited handling when she was a kitten. Many cats hate to be picked up, but if you begin when they are 2 to 7 weeks old, they will quickly become used to it. If Loli is not happy on your lap, give her tidbits to encourage her to come to you. Stroke her and reward her if she stays there calmly. Once she tolerates this, try picking her up just briefly. Put her down as soon as she growls, but if she allows you to pick her up and hold her for a moment, reward her. Work hard at it, and she should become more amenable.

◀ *Eating the flowers from a vase is an annoying habit – and some may be poisonous to the cat. Say 'no' firmly and offer your cat alternative things to chew, such as grass or some dried food.*

Bonding with Your Cat

CONTRARY TO THEIR IMAGE, CATS ARE SOCIAL animals and can form successful attachments to humans and other family pets. Your relationship with your cat depends on its individual temperament, its early socialisation (see pages 44–45), and how you behave with it. Not all cats appreciate close physical contact, which can be disappointing to an owner who wants a cuddly pet. Certain breeds, such as Siamese, tend to be more affectionate; other breeds, such as Abyssinians, are more independent. There are exceptions.

People who don't like cats often maintain that cats bother to seek out affection only when they want to be fed. While it is true that cats are initially drawn to the person who gives them food, a close attachment to their owner is something that runs much deeper. Merely filling up the cat's food bowl is not enough to foster this kind of intense emotional bond. A broad range of interaction is an important way for you and your cat to learn more about each other, so you should play with, talk to, and respond to the cat in as wide a variety of situations as possible when you are first getting to know each other.

This does not mean overwhelming the cat with attention. It is particularly important to avoid this in a rescued cat that has been neglected or abused, or is timid and nervous in temperament. Instead, make yourself generally available, and the cat will come to you in its own time when it feels comfortable and relaxed enough.

A Clingy, Demanding Cat

Some cats want to be with you all the time. They stay in physical contact if possible and are vocal in demanding attention. Many of these cats also knead and suck their owner's clothes when they are being held on the lap. They may show signs of stress if they are left alone during the day or do not get enough attention. In extreme cases this can lead to self-mutilation, fabric eating, and spraying urine indoors (see pages 64–67).

Overly dependent behaviour in a cat can occur if the owner's attitude to the cat encourages it, if the cat needs intensive nursing following illness or injury, and sometimes as a result of behaviour changes in old age. To teach your cat to be more independent, you should discourage its behaviour by not rewarding it for being demanding. Substitute games for petting and prolonged contact. Do not punish the cat for lapses. This is counter-productive and will increase its stress.

Dealing with Aloofness

Underattached cats may be perfectly happy, but their owners may feel rejected and upset. Typically, these cats will not settle on the owner's lap and may run away if they think someone is about to pick them up. The most common cause of this behaviour is lack of early socialisation, but rough owners, traumatic experiences, and invasive handling during illness can also trigger the problem. To promote good socialisation, all kittens should be handled gently on a regular basis, particularly between 2 and 7 weeks of age. If the problem already exists, increase your bond with your cat by rewarding signs of approach and trying to make the cat more dependent. The cat must dictate the pace. Do not pick it up until it is confident and relaxed about being handled.

If Your Cat Is Overattached

Do

- ✓ Limit your contact with the cat by asking someone else to feed, groom and pet it.
- ✓ Ignore any demanding behaviour.
- ✓ Substitute play sessions for cuddling or close physical contact.
- ✓ Stop giving it attention as soon as it starts to suck or knead.

Don't

- ✗ Punish the cat, shout at it, or squirt it with a water pistol.

▲ *Given a choice, your cat will probably want to sleep on your bed, and sleeping right next to you is a sign of a close bond. Whether it is also a sign of overattachment depends on how the cat behaves during the day.*

● ***Will letting my cat sleep with me help us bond more closely?***

If your cat wants to sleep with you, you have already bonded. (Be sure it doesn't have fleas or worms.) If it doesn't want to sleep with you, you won't be able to force it to do so.

● ***My tabby, Miskit, has occasional 'off' days. Is it a bad idea to seek her out and pick her up?***

Any cat that has an off day should be checked by a vet – many cats hide when they are sick. Conditions such as hyperthyroidism and kidney and bladder disease can cause a cat to become irritable and reject physical contact, so be careful how you handle her.

● ***I work from home, and my adult male, Zorba, insists on draping himself across my computer keyboard. What should I do to discourage him?***

Is Zorba getting enough attention? Play with him more and find an alternative place for him when you are working. Settle him on your lap before you start or bring a favourite blanket for him to sit on. A bed that fits on the desk may keep him away from the computer.

If Your Cat Is Underattached

Do

- ✓ Give frequent small meals so that the cat associates your presence with food.
- ✓ Talk to it while you are putting the food out and stroke it gently while it eats.
- ✓ Sit on the floor in its favourite place to encourage it to come over.
- ✓ Offer a game instead of waiting for it to come over for a cuddle.

Don't

- ✗ Chase the cat around the house to make it come to you.
- ✗ Pick it up if it seems at all uncooperative.

Cats and Other Pets

MANY HOUSEHOLDS OWN MORE THAN one kind of pet, and it makes things easier and more pleasant for everyone if the animals live together peacefully. The first-time pet owner is often surprised at how well a cat will get along with some species such as dogs. However, it must be recognised that cats are extremely efficient and active predators, and their basic hunting instincts are triggered by high-pitched squeaks, rustling and tiny, quick movements. It is unwise to trust most cats with pet mice, hamsters – or goldfish.

▲ *A fish in an open bowl is an easy target for a cat. It is probably too much to expect even the most well-behaved cat to resist temptation in these circumstances.*

Early Experiences

As with so many behaviour patterns in all animals, the early experiences of the kitten play a major role in how it reacts to other animals in adulthood. Kittens go through the most sensitive primary phase of socialisation at the age of 2–7 weeks, although the process continues at a lower intensity throughout life, and there is a secondary 'adolescent' phase at 6–12 months.

During these vital early weeks the kitten forms its most intense attachments, for example, to its mother and littermates. It may also develop multiple species identity. In other words, if the kitten is raised with a dog or rabbit at this age, it will assume they are all friends – provided, of course, that there is no outburst of blatant hostility to frighten it. This early conditioning is so strong that if a kitten is raised with a species it normally preys on, such as rats, it will refuse to attack that species even when it is an adult..

As the kitten grows up, this ability to identify with other species rapidly weakens, and thus it becomes much more difficult to encourage a harmonious relationship between the cat and other domestic pets. One striking exception to this rule is a nursing female cat, whose maternal instincts seem to override all others. Lactating queens will suckle rabbits, squirrels and even puppies.

Introducing Cats and Dogs

Dogs and cats should tolerate each other or even develop a close relationship if their introduction is handled carefully. Of course, if the dog has not been socialised with cats, there is a risk that it will regard a cat as small prey to be chased and killed. Some dogs have stronger urges to chase than others, so if you have a dog and are considering getting a cat, you must assess the risks. If you already own a cat and would like to get a dog, choose one that has a placid nature. Avoid ratting breeds such as terriers.

Introducing a kitten to a dog should be a relatively straightforward process:

1. Before you begin, ensure that the dog has had a good walk and a meal so that it is likely to be

fairly content and relaxed. Everyone present should remain calm.
2. Keep the dog on a slack leash and keep the kitten on another person's lap.
3. The dog handler should call the dog to come nearer to the kitten and tell it to sit. When it does, praise the dog or reward it with a tidbit.
4. If the dog seems calm enough, let it sniff the kitten and reward it again for calm behaviour. Repeat this procedure several times daily until both animals are consistently relaxed with each other. You can then allow the kitten to move around the room. Continue to reward the dog for not reacting to this. Never leave them alone together in the first weeks, and keep a trailing leash on the dog initially in case you have to intervene suddenly.

An adult cat's introduction to a dog must be handled more carefully, especially if the cat has not met dogs before or, much worse, has been chased and frightened by them. A mature cat can easily wriggle away from you and flee, and this movement is likely to stimulate the dog to chase. This confirms the cat's fears and perpetuates the problem. You may need to put the cat in a basket well out of the dog's reach for the first few meetings so that the cat can observe the dog without being given the opportunity to escape or defend itself aggressively.

Once both animals seem relaxed, carry on as with a kitten and a dog (above). It may take several weeks before a timid cat can cope with being in the same room as a dog, even if the dog does not do anything to alarm it. Be patient. Take care that the dog does not reinforce the cat's fears by chasing it or barking noisily.

● ***I have a house rabbit called Blackberry and would like to get a cat too. What do you think?***

It's difficult to teach an adult cat to accept a rabbit, so I'd advise you to get a kitten about 6 weeks old or a cat that has lived with rabbits before. Put Blackberry in a carrier so that he cannot run around and excite the cat's instinctive play/chase behaviour, and introduce them to each other for a short while each day. Soon you can let them loose together, but make sure they are always supervised.

● ***Is it any different introducing a cat to a puppy than to a dog?***

Yes – it's much easier. Puppies are unlikely to be aggressive with a cat, and if the cat is a kitten, they will be mutually curious, speeding up the process of acceptance. An adult cat may be apprehensive with a puppy at first but will swiftly gain the upper hand and is likely to remain 'top dog' in their relationship. Keep a close eye on the proceedings – if the puppy intimidates the cat, it may come off worst, with a scratch to the nose or, more seriously, the eyes. Try to keep the puppy calm and relaxed, and introduce it to the cat when it is tired or has a full stomach.

▼ *Cheek by jowl. Deep attachment is possible between a cat and dog living in the same household if they have been raised carefully together or been given time and space to adjust to each other.*

Cats and Children

CHILDREN DERIVE ENORMOUS PLEASURE FROM any pet, and a playful, friendly cat or kitten is ideally suited to interact positively with them. Cat ownership is also an excellent way for children to learn something about the responsibilities and rewards involved in caring for others.

However, children must be taught from an early age to be gentle with a cat and to respect its needs. A pet animal is not a toy but another living being, and rough treatment must never be tolerated. Children should learn that cats, like them, need time on their own and time to rest, and that sometimes it is more considerate of the cat – and equally satisfying for the owner – to watch a cat than to pick it up and fuss over it.

Cats and Infants

An infant and a cat can certainly coexist, but it is best if contact is restricted. Although a human infant is much too large for a cat to mistake for prey, there are other concerns. An infant is not capable of pushing a cat away, and there is a slight danger that the baby might be accidentally suffocated by a cat that snuggles up to it in the crib. For this reason cats should not be left alone with babies. Do not allow your cat to sleep near the baby and place its basket in another room from the baby's crib. If you are pregnant, don't let the cat sleep on the items of baby furniture you bring into the house. Get it used to new sleeping arrangements before the baby is born.

Accustom the cat to a regular routine of feeding, play and grooming – do it now, not after the baby's arrival. Be realistic about how much time you will have for your pet once the baby arrives, and share the responsibility with other family members to prevent the cat from becoming jealous of the infant through neglect. Cats are naturally curious. When the baby comes home, allow the cat to see and smell it – closely supervised, of course. Let it satisfy its curiosity about the baby's accessories, such as nappies and toiletries, before you put them safely away.

● *My 10-year-old daughter has cared for several rabbits and guinea pigs. She would like a pedigree kitten that she can show. Which breed would you recommend?*

She might consider a Ragdoll or Scottish Fold for their placid, mellow nature, or a friendly cat such as a Siamese, Egyptian Mau, or British Blue. It's important to remember that the behaviour of any cat is a product of its environment as well as genes, and good early handling is essential.

● *Our cat Meg is grooming herself so hard that nearly all the hair on her tail is gone. It started when our little boy Daniel began crawling, and she avoids him now. Is there a connection?*

Meg's behaviour could be related to stress arising from Daniel's activities, especially if she was not handled by children as a kitten. Make sure that Daniel is doing nothing to hurt or frighten her, and never leave them alone together. Give her somewhere she can retreat to away from Daniel. I would also advise you to visit the vet for a full assessment and to see if antianxiety medication might be suitable in Meg's case. There is a chance that an allergic reaction to baby products is responsible for her excessive grooming.

● *We recently adopted a rescued British Cream called Henry. He is fine with my wife and me, but although he will get on our daughter's lap willingly, after a short while he jumps off and hisses at her from across the room. Is she doing anything wrong?*

It sounds as though Henry has been roughly handled by a child in the past and he loses his nerve when he is at his most relaxed and vulnerable. Continue to encourage him to go to your daughter for short petting sessions, and get her to take over the feeding and grooming if he enjoys it. In time the trust will grow, but she will have to be patient.

▲ *Children are fascinated by kittens, and a kitten fully returns the love of a child who treats it well. The child must be old enough to respect the kitten's needs, and the kitten should have been introduced to children early.*

The Importance of Socialisation

Not all cats mix well with children. A common cause of problems is lack of socialisation in the cat's early life (see pages 44–45). If a cat is to live with small children, it should ideally have been brought up in a noisy, active family environment and been handled regularly by children when it was between the ages of 2 and 7 weeks. Poorly socialised cats tend to be timid. They will either avoid the children or scratch and bite them if picked up against their will. Many families would rather have a kitten than an adult cat, and this usually works out satisfactorily, as long as the kitten has been accustomed to being lifted gently from an early age. Children simply love to pick up and cuddle tiny animals.

Toddlers should always be closely supervised with cats to minimise the danger to both. Parents must teach their children not to shout at the cat, pull its tail, drop it, or squeeze it – or to lock it in a cupboard, the washing machine, or worse still, the microwave. The unwanted attentions or violence of children can exacerbate stress-related behaviour even in well-socialised cats.

Conversely, problems with poorly socialised cats can arise even when a child is doing nothing wrong. If a cat is inclined to show predatory behaviour towards people's moving limbs, there is a great likelihood that a child's feet or hands will be attacked. Any sign of this behaviour is a problem and should be tackled early (see Aggression Towards People, pages 62–63). If you are having any difficulty, talk to your vet.

The Importance of Play

THE PLAYFUL NATURE OF THE DOMESTIC cat is one of its most endearing traits. Owners with two or more cats get endless pleasure from watching them chase, stalk and wrestle with each other. Kittens' play helps them improve strength and coordination, investigate each other and practice adult behaviour and survival skills. Adults' play is an outlet for energy and reinforces natural behaviour.

▲ *Catch me if you can. Two littermates enjoy an energetic game of chase across a flower pot. Play is an essential part of a kitten's training for adult life.*

Kittens at Play

Kittens begin to play almost as soon as they open their eyes and move around (10–14 days). Initially, their play consists of clumsy, uncoordinated attempts to pounce, wrestle and roll over each other, with a certain amount of mock biting, paw batting, and scratching. As the kittens approach 8 weeks of age their interest in toys increases, and they will chase and pounce on anything that moves.

This kind of play practises and develops the kitten's predatory skills: stalking, chasing, pouncing on the prey, and 'killing' it. It continues into adult life: think how quickly a cat's attention is drawn to another cat's flicking tail, a dangling piece of string, or a passing butterfly. However, kittens that are deprived of play may still become habitual hunters, and cats that do not hunt will still play. Encourage the habit by giving your kitten frequent play sessions.

▼ *A wide variety of toys are available for kittens and cats to play with on their own or in an interactive game. Bells are fascinating, but must be securely attached.*

Safe and Simple Toys

With a little ingenuity you can invent some simple toys and games to keep your cat amused.

- Cut holes in the side of a large cardboard box. The cat can jump in and out, and hide inside.
- Thread a small ball, bottle cork, or empty spool firmly on to strong string. Make sure the object is too big for the cat to swallow.
- Open up sheets of newspaper on the floor for the cat to tunnel through or give it rolled-up newspaper balls to play with.
- Swing the beam from a torch back and forth across the floor for the cat to chase.

● *Is a kitten happier playing by itself, with a person, or with another kitten or cat?*

Kittens living together or with adult cats tend to be more sociable with each other, while single cats are more likely to interact with humans. If you want a pet who is deeply attached to you, you should get a single kitten.

● *How often should I play with my cat to prevent him from becoming bored and restless?*

A kitten needs frequent sessions of 15 minutes or longer; this will aid its physical, mental and social development. An adult cat is fine with ten minutes twice a day, as long as it has other ways to amuse itself.

● *Why does my cat Lenka sometimes bite or scratch me when we are playing? She was an orphan kitten, hand-reared from the age of 2 weeks.*

This sounds like play aggression, which is most common in hand-reared kittens. Keep Lenka's play sessions short, reward good behaviour with stroking or food, and stop immediately she shows any warning signs of aggression: dilated pupils, tail twitching, ears going back, stiffening of the body, or unsheathing of claws. Introduce something like a ball on a string to play with rather than using your hands directly in play.

● *Nimrod plays with all his toys except his catnip mouse. Why does he ignore it?*

Catnip (a strong-scented, dried mint) contains nepetalactone. Some cats react strongly to it, and it can induce a drug-induced state – they may even appear to hallucinate. However, studies indicate that a little less than half of all cats show little or no response to catnip, so your cat is by no means alone in ignoring his toy mouse.

Keeping Your Cat Amused

Provide your cat with plenty of toys that you can play with together. This will help develop your relationship, act as an outlet for your pet's energies, and prevent it from becoming bored, lazy and obese, especially if it is an indoor cat. Suitable toys are anything that the cat finds interesting, provided it is safe. Beware of plastic bags that may asphyxiate the cat, full spools of thread or balls of wool that can be chewed and swallowed, small buttons and other detachable parts.

An indoor cat activity centre is an excellent way of providing exercise and entertainment. It can be built of boxes and surfaces at different levels, with perching areas, ropes to climb and swing on, tunnels to chase through and hide in, and moving toys. Incorporate a scratching post or natural bark in the design. A large room is not necessary, just vertical space (about 2m/6ft) and plenty of imagination. Try changing the layout occasionally to give the cat variety.

▶ *A swinging toy allows a kitten to make practice feints before it springs and seizes the object. Through regular play sessions, this young owner is building a lasting relationship of affection and trust with her pet.*

Hunting

It's easy to forget that cats were first domesticated to hunt small vermin. Their physical attributes contribute to their efficiency as predators: large, forward-facing eyes for a good field of vision; ears that perceive high-frequency sounds well beyond the human range (such as those emitted by small rodents); extraordinary agility and balance for leaping and pouncing; and deadly claws and teeth for grasping and killing prey. When the final bite is delivered, it will sever the victim's spinal cord between the neck vertebrae, killing it almost instantaneously.

Cat owners are often unhappy about the hunting tendencies of their pets. Many of us would be content for them to kill rats and mice, but most of us dislike the bird-hunting habit. Fortunately, cats chase and stalk birds far more frequently than they capture them. Those that are caught are likely to be young ones just leaving the nest, particularly ground-nesting species. One way to prevent this is to put a bell on your cat's collar, so that it rings when the cat moves, thereby alerting any birds nearby. (Some cats, however, eventually become adept at moving without shaking the bell.) Hunting cats will also catch beetles and other insects, reptiles, small rodents, larger rats, and even rabbits, according to availability. Small prey are eaten head first, but the intestines are often spurned. Prey is not always consumed on the spot but may be carried back – sometimes still alive and struggling – as a gift for the owner or other cats. In your cat's eyes, this behaviour is perfectly natural and right, however distasteful you may find it yourself.

▼ *Predatory behaviour is instinctive in cats and can be seen from the time a kitten is 5 weeks old. Most of their prey will be mice, which are easier to catch than birds.*

How Cats Learn to Hunt

Hunting is partly instinctive and partly a skill that the kitten learns from the mother. The hunting, fighting and feeding activities are controlled by different areas of the brain. Well-fed cats and neutered toms may be equally as skilled and dedicated at hunting as unneutered toms.

A hunting mother introduces prey to her litter from the age of about 5 weeks. Initially the prey is dismembered for the kittens to try a morsel as an alternative to milk. Gradually, however, the kittens learn to play with the unfortunate captive before it is killed and eaten. There are several theories to explain why adult cats indulge in this apparently cruel activity. It is thought by some to be a play sequence that perfects the cat's hunting technique, but others argue that at the moment of kill the cat is in conflict about which of two different instincts to obey: whether to kill or to flee. The cat's behaviour, as it appears to advance and then retreat, resembles play, but it may be a survival technique – when facing large prey, a moment's misjudgment by the cat could lead to injury.

Practice hunting is an important element of kitten play. Littermates will begin to stalk each other at 3 weeks old. By the time they are 5 weeks, they have developed the three basic hunting techniques of pounce, swat and scoop.

▼ *Standing on the hindlegs frees both front paws to swipe at birds. Don't hang your bird feeder low enough for your cat to reach it. A collar with a bell would alert the birds that a cat is nearby.*

● ***How can I find a cat that won't hunt the birds in my garden?***

All cats are predatory by nature. If you want to increase your chances of preserving bird life, select a kitten from a nonhunting queen, and bring it home when it is 6 weeks old. Stimulate it with a wide variety of toys, or get another kitten so that they can amuse each other. Keep the kitten(s) well fed. This will not remove the desire to hunt but may reduce the desire to kill and eat the prey.

● ***We have just 'inherited' two country-bred adult females. They bring a steady stream of small dead mice and birds into the house. What can we do?***

Try to catch the cats as they bring their victims into the house and spray them with a water pistol or make a loud noise to startle them. This will only work if you catch them in the act; once the bodies have been left on the floor, it is too late. After a few occasions they should get the message that dead bodies are not welcome indoors. Be affectionate to them at all other times, however, so that they do not decide they themselves are not welcome either.

● ***If my cat hunts birds and mice regularly, do I still need to feed him a normal diet?***

Certainly your cat could survive on hunting, as long as the prey was plentiful and competition in the area not too great. However, if you do not feed him at all, he may drift away to better hunting grounds or more generous neighbours. Also, the danger of letting your cat eat its fill of prey is that it may pick up tapeworm or even poisons from birds and rodents. Don't drive your cat to hunt for food by letting him go hungry.

▼ *Toying with its prey may be the cat's way of practicing its technique. Humans may consider this cruel, but like all hunting behaviour in cats, it is completely natural.*

Territorial Behaviour

CATS ARE FULL OF CONTRADICTIONS. THOUGH domesticated, they may still return to the wild. They coexist in feral and wild colonies, but will refuse to share a small patch of lawn with the cat from next door or a favoured chair with a sibling. They will attach themselves to you as long as you feed them to their liking, but can prove fickle if there is better to be had elsewhere. Given their own garden, many of them prefer to climb over the fence and relieve themselves next door.

Two Key Resources

Space and food are the most important resources in a cat's life. Retaining the powerful instincts of the hunter in the wild, cats are highly sensitive to anything that may restrict their access to either. A cat mostly acts to control its territory, which it has chosen (or accepted) because it is a reliable source of food. Another cat that enters your cat's territory is likely to be perceived as a threat, and the rival must be warned off or, as a last resort, fought off to preserve your cat's peace of mind. If the food supply at home is not reliable, your cat will enlarge its territory – that is, go roaming or adopt your neighbours.

● ***My neighbours are threatening me if my cats don't stop relieving themselves in their flower-beds. Is there anything I can do?***

If your neighbours get really angry, they may start intimidating your cats by shouting at them or throwing things. A better approach is to provide an irresistible toileting area in your own garden, such as a large sand pit that can be cleaned out at intervals. Cats find the soft texture of sand very inviting.

● ***My tomcat, Rocky, was neutered early but he still beats up the neighbours' cats. The vet's bills are getting expensive, and my neighbours are angry. Do I have to get rid of him?***

This is a common problem but not always easy to solve. You can try keeping Rocky indoors, or set up a schedule so that he is only allowed out at prearranged times when the other cats are kept in. You may also wish to find a pet behaviour specialist to help Rocky. Medication may reduce aggression temporarily, but a programme to prevent him attacking other cats is still necessary. You may need to find ways of redirecting his energy in active play, for example.

● ***My two neutered females are from the same litter, but Carmen has gradually assumed sole possession of the living room, and Tosca won't go in there at all now. Is something wrong?***

Sibling pairs that have been raised together usually get along fine. If they are friendly in all other parts of the house, then I suspect something unpleasant has happened to Tosca in the living room. It may or may not involve Carmen, but try coaxing her in when Carmen is not there. If she seems tense, stroke her to calm her down. Feed her in the room for a few weeks to develop pleasant associations. If the cats are not friendly elsewhere in the house, the problem is their relationship, and you will have to work to improve that.

◀ *Actual fighting is the last resort in territorial disputes. Cats other than intact toms tend to avoid each other's private space, and usually a display of bravado alone will deter intruders.*

▲ *Like their wild ancestors, domestic cats feel more secure if they can survey the scene from on high. Social status is flexible in cats, and the cat that occupies the highest spot has an advantage, if only temporarily.*

If cats were not so adaptable, the presence of domestic pets in a suburban neighbourhood would lead to endless fights, but if the area is not overcrowded and all the cats are adequately fed at home, most of them will coexist more or less peacefully. Cats with access to the outdoors tend to carve out an area around their owner's house and garden. This may be shared with other cats in the same household and overlap with that of neighbouring cats. Some cats are adept at 'time-sharing': one cat may use a certain wall to doze on in the morning, while another may claim it in the afternoon. Hunting cats are more aggressive in defence of their territory at dawn and dusk, when small rodents are more active.

Hormones and Territory

Territorial aggression naturally increases when two cats are competing for scarce resources. Hormones also play a role. Unneutered males need more space to control, hence their tendency to get into regular fights with other males whose patch they trespass on. Nursing queens do not demand as much space but will defend their patch vigorously. Neutered cats of both sexes are the least territorial of all and engage in more friendly body contact with other cats. This is why neutering is an effective way of reducing fighting in males (see pages 74–75).

If a territory is large, your cat cannot be everywhere at once, so it leaves reminders of its presence by frequent patrolling and urine spraying of the boundaries. These scent marks tell other cats the sex and hormone status of the cat that left them, as well as its route and the time of its visit. Trees and fences are clawed to leave visual and scent marks, and sometimes faeces are left uncovered as an added warning. Like spraying, this is normal behaviour for a cat, though it becomes a problem when it occurs indoors (see pages 64–65) or in your neighbour's garden.

Clawing

Claws are vital assets for hunting, balancing, climbing and in self-defence. When not in use, the cat's claws retract into a protective sheath so that they are hardly visible, but when they are needed for action, tiny individual muscles propel them out of the sheath to the tip of the toe. The sharp tips dig into surfaces when the cat is climbing, and the dew claws (the extra claw on the inside of the leg) act as side grapples.

When the cat is hunting, the front claws are used to grip the prey; the dew claws and hind claws come into action only if the cat is dealing with very mobile or larger victims. In a fight, the claws of all four feet pierce the skin, often inflicting deep puncture wounds. Those on the hind feet are used to deliver rapid, alternating, slicing cuts to the prey's soft underparts.

The cat keeps its claws in good condition by scratching on rough surfaces such as trees or fenceposts. Scratching keeps the nailbeds clean of dead skin and debris and allows the outer layer of dead claw to be shed. This exposes a new pointed tip underneath. The discarded outer layers are often found embedded in the surface of scratching posts. Scratching also serves another important function – it enables the cat to identify its territory with a combined visual and scent mark. The scent is derived from glands on either side of the pads. Usually a cat's regular scratching posts are outdoors in areas where a number of cats are competing for territory, but some cats persistently scratch indoors.

Protecting Your Furniture

Kittens practise the art of claw maintenance from an early age. Since they have to be kept inside at least until their vaccinations are completed, and many owners delay a few more weeks until the kitten is big enough to fend for itself in the outside world, they are more than likely to get into the habit of scratching furniture and carpets. Avoid this by providing a substitute object for your kitten to use such as a scratching post bought from a pet shop – or make one yourself by wrapping tough cord around a piece of wood. Encourage your kitten to scratch it. If you catch the kitten in the act of scratching the furniture, hiss at it loudly or clap your hands sharply – this is usually enough to stop the activity. Do not punish it. You can also ask your vet to trim your kitten's claw tips. This does not stop the habit, but it does minimise damage.

▶ *Given free access to the outdoors, cats naturally prefer to scratch on trees or wooden fenceposts. It is particularly important to provide indoor cats with suitable alternatives to condition their claws.*

◀ *Fringes on soft furnishings are particularly tempting playthings. An 8-week-old kitten, when left unsupervised, can shred fabric like this in a matter of minutes.*

▼ *A scratching post can provide different textured surfaces and a toy to play with. Some posts also incorporate grooming brushes.*

Toy with a bell

Column covered with sisal rope

Sturdy base covered in heavy-duty carpeting

By providing an attractive alternative, it is possible to train an adult cat to turn its attentions away from your furniture. Watch to see what objects it scratches against outdoors and make a scratching object from the same material. For example, if it likes the bark of trees, place a piece of treebark directly in front of your cat's favourite indoor scratching area. Once the cat is in the habit of scratching it, slowly move the substitute to a more suitable position away from the furniture. Every time you catch it scratching in the wrong place, utter a sharp 'no!' and redirect its attention to the desired object.

● *A friend has told me that a scratch from my cat can make me sick. Is this true?*

Yes. If a scratch wound from a cat becomes infected, it may induce a condition known as 'cat scratch disease' in humans. It was once thought to be passed on by bacterial organisms found in the nailbeds of cats, but the bacteria are now thought to be present in the mouth and transferred to the claws when the cat grooms itself. Clean all bite and scratch wounds carefully with antiseptic. If swelling or pain develop, consult your doctor.

● *What is your view of the practice of declawing cats to prevent them from destroying the furniture?*

Declawing is highly controversial. Many vets believe that it is a cruel and unnecessary act of mutilation, and their professional associations do not support the practice. Some national cat associations do not allow declawed cats to enter shows. The practice is most widespread in the US, where a high proportion of cats are kept indoors: discuss it with your vet when you acquire your kitten. You can also try fixing nail caps – plastic tips – onto the claws to reduce damage. But regular claw trimming, together with the provision of a substitute scratching post, is a much kinder way of minimising the scratching problem.

Aggression Towards Other Cats

CAT OWNERS TEND TO ASSUME THAT FUR WILL always fly if cats do not like each other, but it is unusual for cats to hate each other so much that they attack on sight. The signs of social discord are typically more subtle than this. The relationship between cats can be assessed by watching their body language and posturing, particularly tail and ear positioning and whether or not they greet each other, groom and rub each other, or sleep next to or separately from each other. An aggressor will approach another cat and stare at it, while the other looks away or retreats. If one of the cats refuses to back down, they may both begin growling, hissing, spitting, striking out and biting. Observe your cat's behaviour with other cats. If it is normally placid but suddenly starts lashing out, you should take it to the vet: an underlying illness could be responsible for this sudden display of aggression.

Aggression towards Strange Cats

The most common cause of problems between cats that do not live in the same household is conflict over territory. This is particularly widespread in urban areas where many cats' personal domains overlap. A cat allowed to roam freely outdoors in a typical suburban neighbourhood is bound to cross another cat's territory. Similarly, it is difficult to keep other people's cats out of your garden – high fences pose little challenge to most agile cats.

There is probably little you can do to prevent your cat from being drawn into territorial disagreements with other cats, aside from having it neutered at puberty – particularly if it is a tom (see pages 74–75). However, if your cat is frequently on the receiving end of aggression, you may decide to keep it indoors as a precautionary measure. It is extremely important that you keep your cat's vaccinations up to date, as infectious diseases are readily transmitted by a bite from an infected cat.

▼ *A full-fledged fight is developing. The cat on the left is in a dominant rearing stance with its ears forward, showing that it is the aggressor, while the cat on the right defends itself from a crouch, its ears laid back.*

▶ *Competition over food can lead to aggression. Each cat should have its own food dish, but if one is already dominant, it may well finish its own meal and then expect to take over the other's. In this case, you will have to feed them separately and prevent interference.*

Aggression in the Same Household

Kittens that are raised together usually get along well as adults, whether or not they are siblings. Parents and kittens generally remain affectionate to each other as well, but there may be exceptions. Disputes over hierarchy can occur if two individuals perceive each other to be of similar status. Disputes sometimes arise when one cat in a household reaches puberty (6–10 months) or social maturity (2–4 years), or when a newcomer is introduced to an established group. Overcrowding and stress can cause breakdown in social harmony, but sometimes there is no explanation for sudden enmity. In severe cases, rehousing one of the cats is the only solution.

When introducing a new cat to your household, try to minimise potential friction by following the advice on pages 48–49. Smell is the key to familiarity between cats, so it is important to allow at least one week for resident cat(s) to get used to a newcomer's scent. The new cat should not be given freedom of the house until it has been accepted by the established resident(s). It may be some weeks before you can tell if they are going to get along. They may never become the best of friends, but this need not be a cause for concern as long as they respect each other's space.

● ***My cat Toffee has begun hissing at her sister, Fudge, who just came back from a week in the vet's hospital. What is wrong?***

Fudge probably smells different to Toffee, with the smell of the hospital still clinging to her. Bathe her to remove the strange odours or, better still, towel off both cats – this should help to mix up their smells and reconcile Toffee to her sister. If this doesn't work, separate them to prevent confrontations and seek expert advice.

● ***Our cat Zac was watching the tom next door through a window, growling at it. When Simi, our other cat, walked by, Zac suddenly bit her. Now Simi is terrified of him. Is this likely to happen again?***

Zac wanted to beat up the neighbour's cat, and Simi came along at the wrong moment. This is called redirected aggression, and may sometimes be aimed against humans. Try to prevent Zac from seeing the real object of his hatred and intervene instantly with a water pistol if he shows any aggression to Simi. Allow them plenty of space, but aim to feed them side by side and groom them in the same room. However, don't leave them together unsupervised for now: put a different bell on each cat so you know where they are.

● ***The local tom chases our two neutered females indoors and has begun eating their food and – worst of all – spraying in our house. What can we do?***

The best solution is probably a pet door with a device that recognises sensors in special collars worn by your cats so that the door can only be opened by them. Also, try talking to the tom's owner; or if it appears to be a stray, call the local animal shelter.

Aggression Towards People

AGGRESSIVE CATS ARE NOT AS DANGEROUS – OR as common – as aggressive dogs, but no owner enjoys being bitten or scratched by their pet, or wishes to expose a child to injury and distress. Although cat bites and scratches are usually less harmful than injuries from dogs, they can still cause disease. Wounds sometimes turn septic or are infected with the Bartonella hensalae organism, responsible for 'cat scratch fever'. This infection usually leads to nothing more serious than enlarged lymph nodes in someone who is healthy, but in anyone whose immune system is already compromised, such as a person who is receiving chemotherapy, it can cause severe illness. In the United States there are more than 20,000 cases of cat scratch disease annually.

Why Cats Attack People

Pain is very often what makes a normally good-natured cat turn aggressive. If your cat suddenly attacks you for no apparent reason, check it for signs of injury. Have it thoroughly examined by your vet to rule out any underlying illness such as rabies, urinary disease, or hyperthyroidism. Then look elsewhere for the cause. A frequent reason for aggression is fear. This is most likely to be seen in a cat that has been poorly socialised (see pages 44–45 and 82–85), has suffered cruelty at some stage in its life, or has come to associate a specific event, such as a visit to the vet, with distress and fear. The body language of these cats (see pages 42–43) will show that they are intimidated and fearful. They will attack if they are pushed too far.

Play aggression is normally directed towards family members and occurs most frequently in cats that have been hand-reared or weaned too early. These kittens, deprived of normal socialisation, fail to learn to inhibit their bite or sheath their claws during play, as they would if growing up with their mother and littermates. A mother cat instantly reprimands any kitten that is getting too rough, whereas without the benefit of this discipline, the kitten does what it likes. This behaviour begins when the kitten is playing with humans, and if its basic temperament is friendly, owners may let the attacks go uncorrected.

Predatory aggression is normal when directed towards rodents and birds, but some cats may begin stalking and lunging at their owners' feet or hands. If the target is a child or an infant, this behaviour is dangerous and must be prevented, even if it means confining or harnessing the cat until it is under control. Use startle techniques to interrupt the predatory stalk (see pages 44–45), and fit a bell onto a breakaway collar to help you be aware of the cat's movements and anticipate an imminent attack.

Sometimes a cat will attack its owner if it is unable to get near another cat that it can see or smell (redirected aggression). Occasionally, during a petting or grooming session, a cat may lash out at its owner. This often occurs in a cat whose aggression is related to status (rarer in cats than dogs). These cats will solicit attention at inconvenient times, when their owners are talking on

If Your Cat Is Aggressive to People

Do

- ✓ Take it to the vet to rule out physical causes.
- ✓ Reduce physical contact with the cat.
- ✓ Startle it with a sudden noise or a water pistol if you catch it attacking.
- ✓ Divert aggressive play to a toy held at a distance away from your body.
- ✓ Confine or harness the cat if a small child is at risk.
- ✓ Clean any scratches or bites with antiseptic.
- ✓ Consult a behaviour counselor.

Don't

- ✗ Lock eyes with an aggressive cat.
- ✗ Use physical punishment.

▲ *Extreme aggression such as this is rarely directed at owners unless the cat is in fear or pain. If there is no obvious explanation, have your cat checked by your vet as soon as possible to rule out injury or illness.*

the phone, for example, and may block corridors or narrow entrances when people wish to go past. Some cats will stare at their owners to try to intimidate them (see pages 42–43). Professional help is required to treat the problem, and usually involves reducing physical interaction, avoiding confrontational situations, and reacting rapidly to signs of impending aggression. There should be no physical punishment as this is likely to increase the cat's aggression. In some cases a vet may prescribe psychoactive drugs to keep the cat calm while it is being treated.

Q&A...

● ***We have had Nijinsky, our 1-year-old Siamese, ever since he was 12 weeks old. He is very friendly but tends to play rough, sometimes biting and scratching. My 5-year-old son loves to play with Nijinsky, but I am concerned that he will end up being injured.***

This behaviour is sometimes seen in bold cats whose owners have inadvertently encouraged aggressive play. Introduce 'hands off' games with your son by directing Nijinsky's attention to rolling objects or toys on the end of a long string. The toys should be held well away from your son's body. If Nijinsky shows signs of imminent attack, startle him with a sharp hiss or loud noise and immediately direct his attention to an appropriate toy. Make sure that he has plenty of exercise and activity to help him work off his energy.

Spraying and Soiling Indoors

SOILING OR URINATING INDOORS IS BY FAR THE most common reason that cats are given away or euthanised. If your cat has this problem, you will want to deal with it as soon as possible, if only to keep your house free of unpleasant odours and save yourself cleaning chores: once the habit becomes established it will be difficult to break. Do not assume that your cat is being lazy or obstinate if it fails to use its litter box; it may be a sign of an underlying health or behaviour problem. Punishing your cat will have no effect except perhaps to make the problem worse.

Possible Causes

Examination by your vet is essential to establish whether your cat has a medical condition. Kidney and liver disease, cystitis, diabetes mellitus, spinal or pelvic injury, hyperthyroidism, arthritis and treatment with steroids can all cause abnormal urination; bowel problems may be due

▲ *A cat will spray against an upright surface indoors if it feels threatened by an intruder. Telltale signs are thin wet patches on the wall or furniture. A puddle on the floor is more likely to indicate an urinary problem.*

● ***Anais, our shorthair, has started to mess in the living room near the door to the kitchen. It began after a neighbour's cat got into the house through the pet door and sprayed in the kitchen. What can we do to cure it?***

First, lock the pet door so that the visitor cannot return. If your back door is glass, cover it with a curtain so that the intruder can't stare at her through it. Don't let Anais into the living room. Use hot water and a biological odour eliminator to clean the soiled areas. Make sure that Anais' litter box is in a quiet place – nowhere near the pet door. If she tries to defecate elsewhere, confine her to one room for a few days with access to a litter box and water. Feed her regularly and make sure she has plenty of attention. When you let her back into the living room, put her dish over the previously soiled area. You may wish to install a magnetic flap, activated by a device on Anais' collar, on your pet door.

● ***Tortie, our 15-year-old cat, has started to use the bath and basin as a toilet. She sleeps upstairs, but her litter box has always been in our kitchen. Is this behaviour a sign that she is getting senile?***

The litter tray is probably too far away for Tortie to get to quickly now that she is getting older. Have her checked by the vet for arthritis or any other condition that might cause pain or stiffness. Try putting a litter box upstairs near to Tortie's sleeping spot (but not too close). Keep the bottom of the bath and basin filled with a little water to discourage her from using them.

● ***A month ago we moved to a new house. Since then Raja, our neutered Oriental Shorthair, has started spraying indoors. His first target was some boxes I had moved in from the garage. What do you think is making him do this?***

A neighbourhood cat may have wandered into the garage and sprayed on the boxes. Get rid of them and clean all soiled areas with hot water followed by a surgical spirit spray. You should then cover the area with a pheromone spray as this will make Raja feel more secure. Make a lot of fuss over him. If he continues to spray, consult a pet behaviour counselor. It may be necessary to confine Raja to a restricted area for a time. Antianxiety medication can also help.

to a virus or bacterial infection, parasites, colitis, or even sensitivity to certain foods.

If medical causes are ruled out, the answer may lie with the litter box itself. Some cats will not use their box if it is in the wrong place. Don't position it next to where you feed the cat – this is almost guaranteed to cause problems. Many cats hate being watched while they relieve themselves: is the box too public? Perhaps you have recently changed the type of litter for one it dislikes (see pages 20–21). Has the cat been startled while using the box, and developed an aversion to it as a result? Finding the root of the problem takes time and careful observation.

Territorial Marking

Cats' urine and faeces are used as social signals to other cats (see pages 42–43) and to surround the cat itself with its own reassuring smell. If the cat does this indoors, it is probably feeling threatened. This type of marking tends to be found around exits such as doors and windows. Usually the cat sprays against an upright surface rather than on the floor, leaving a thin wet patch on a curtain or wall, but occasionally it may leave a puddle as well. If your cat leaves small stools strategically positioned on the furniture, it may be showing extreme insecurity. It will take more time to cure – seek professional help.

Breaking the Habit

Once a cat has singled out one spot to relieve itself in, it will keep coming back to the same place because it has the right smell. Take the following action to prevent this happening:

1. Immediately clean the soiled area(s) with a biological odour eliminator. Ordinary disinfectants will not stop the smell; scented ones may encourage the cat to mark the spot again.

2. Place the cat's food bowl over the spot. Cats avoid soiling where they eat, and this should be an effective deterrent.

3. Make sure that the cat's litter box is in an acceptable place, contains the right kind of litter and is cleaned regularly.

4. If the problem persists, confine the cat to one room or an indoor pen with only its bed, water and a litter box. Allow it out to eat and for plenty of play and cuddling sessions. Gradually let the cat into the rest of the house but keep a close eye on it. As soon as it shows any sign of urinating or defecating in the wrong place, return it to the confinement area and begin again.

5. If all else fails, you may need to seek professional help for your cat.

▼ *After an accident, remove all traces of odour with hot water and a biological cleaner. This will stop the cat returning to the spot.*

Stress and Phobias

EVEN A HEALTHY, CONTENTED CAT MAY SUFFER from stress at some stage in its life. Moving to a new house, the arrival of a baby in the family, or the appearance of a new pet are the most common events that affect a cat's environment. How quickly your cat adjusts in these circumstances depends partly on your behaviour and partly on its temperament, which is determined by a combination of genetic and environmental factors. Broadly speaking, cats have one of three kinds of personality: bold and friendly, active and aggressive, or timid and nervous. The fearful personality is most strongly influenced by heredity and the most likely to have difficulty adjusting to new situations. It will show the most exaggerated physiological and psychological responses to any unpleasant or unfamiliar event. Such events can range from hyperventilating to hiding under the bed, refusing to eat, and eliminating in the house (see pages 64–65). If the stress response continues over a length of time, it is called a disorder and may become a permanent condition.

Some cats may suddenly develop a full-blown phobia – an extreme, apparently irrational fear of certain events, places, or even people. The cat may seem calm until confronted with the object of its dislike, when it suddenly becomes hysterical. Children, dogs, the cat's carrier, and loud noises such as fireworks and thunderstorms are possible triggers of phobic reactions. Gradual controlled exposure should desensitise your cat to people, animals and its carrier – even the vacuum cleaner – but unpredictable loud noises are more difficult to overcome. Your vet should be able to prescribe medication to soothe the cat and may recommend a behaviour expert.

▲ *Most children love cats, but a child's attention can overwhelm a kitten or timid cat. A kitten that is squeezed hard may grow up to be afraid of children.*

Common Stress Disorders

Stress may be expressed in many different ways, but one of the most common ones is that the cat stops eating. If this happens, your vet must check it for a medical problem before looking at any environmental cause. To encourage your cat to eat, offer its favourite foods, feed it from your hand, then gradually drop bits of food into the bowl. Stroke and praise the cat while it eats. Next, fill the bowl but put it on your lap so you can hold the cat while it eats. Then put the bowl back on the floor but continue stroking and praising the cat while it eats. Your vet may prescribe an appetite stimulant to encourage eating.

Fabric eating is another sign that your cat may be upset. Wool, string, plastic, even rubber or electric cables may be eaten by an anxious cat. This is most common in Siamese and Burmese and in cats weaned too early (2–4 weeks). It often begins in adolescence but can also be a reaction to stress. While you look for the cause, prevent access to the material and provide dry biscuits or large bones with sinew attached.

● ***Are pedigree cats more prone to stress and phobias?***

...Orientals tend to be the most reactive, but on the whole individual temperament is more important than the breed. In addition to heredity, the way a cat responds to stress is influenced by the mother's behaviour, the kitten's early socialisation, the home environment and the owner's behaviour. This is true for all cats, pedigree or not.

● ***I am 6 months pregnant, and my doctor has told me not to handle my cat Shengo, who has now developed a bald area on his tail which he licks constantly. Is he upset with me?***

This could well be a reaction to stress, caused by your distancing yourself from him. However, Shengo may have a skin problem, so have your vet examine him. There is no danger in your handling Shengo himself, only in handling his litter (see pages 86–87). Delegate this chore to someone else and pay attention to Shengo, keeping in mind how much time you will have for him when the baby arrives (see pages 50–51). Perhaps your partner could help by feeding and grooming him more than you do.

● ***Isolde, my Siamese, continually pulls threads out of my sweaters and chews them. Apart from the nuisance, I'm afraid it might harm her. Will it?***

Fabric eating can lead to physical harm if it causes an internal obstruction, and a cat that chews through wiring may be electrocuted. You should gently discourage the habit if your cat has it.

Cats are normally very clean, but a stressed cat may spend up to half its time grooming itself, leading to bald patches (alopecia) and even mutilated skin. If your vet rules out an underlying physical problem, medication may be prescribed to keep the cat keep calm while you eliminate potential sources of stress.

Some cats bite their tails, rush around spasmodically and roll on the floor, and even mutilate themselves (this is not the same as obsessive grooming resulting in skin damage). Their skin may twitch, especially along the back. These twitches can be signs of a flea allergy or back pain, or be due to stress. Take the cat for a check-up to find the cause. Anti-anxiety drugs may be particularly useful for this condition.

Plant eating is common in indoor cats who do not have enough to keep them amused (see pages 30–31 and 52–53). Provide your indoor cat with grass or herbs to chew and be sure that it has plenty of toys and something to climb on.

Finally, it can help an anxious cat if you provide it with a safe place where it can hide when it feels threatened. The warmth and quiet of a linen cupboard is a favourite place to retreat, and a shelf or high furniture from which it can survey the world may make it feel more secure.

▶ *A timid cat that cringes at strange noises or hides from people may be prone to stress disorders. Be aware of sources of stress in your household.*

Your Cat's Health

HOWEVER MUCH CARE YOU GIVE IT, YOUR CAT is likely to suffer at least one illness during its life. Early recognition of a problem by the owner and prompt veterinary attention are the keys to good health and a long, active and happy life for your cat. As a cat cannot explain its symptoms, it is important that you take note of any physical or behavioural changes that occur and describe these accurately to the veterinarian. To help you, and to give you a better understanding of your cat's health, this section starts by describing the cat's basic anatomy. It then lists the most common infectious diseases and parasites of the cat.

The section concentrates on the medical problems that owners are most likely to encounter. The starting point is the particular symptom, or group of symptoms, that alerts their concern to something being wrong. All the likely causes of that condition are discussed, followed by advice on prevention and possible courses of treatment. You'll also find advice on whether to have your cat neutered, on breeding, and on first aid.

Getting the Most from Your Vet

YOUR VET IS AN IMPORTANT PARTNER IN CARING for your cat. Regular checkups allow both of you to spot any problems before they become serious, and small changes that you have not noticed can be obvious to the vet. Regular visits also mean that your vet gets to know you both. If possible, try to see the same vet each time.

It is a good idea to register with the vet before your cat needs treatment. This allows you to find the practice and meet the staff under relaxed circumstances, not in an emergency situation. Also, many good flea-control and worming drugs may be dispensed only to registered clients whose cats are under the direct care of the vet, so an examination is usually required before any medicine can be dispensed.

Choosing Your Vet

Take some time over your choice of vet, just as you would your own doctor or dentist. You and your cat should both like and trust the vet, who should be gentle as well as expert. Asking a cat-owning friend or neighbour for a recommendation is a good place to start.

A number of factors have to be taken into consideration when choosing your vet. Distance is an important one; many cats become anxious when they travel very far, and a long drive is a serious disadvantage in an emergency. If you do not have a car, the vet should be accessible by public transportation, and if you work during the day, you will want to find a practice that is open at least one evening or on Saturdays. Aside from these practical considerations, you may wish to look for a veterinary hospital that provides a cats-only clinic. This has the advantage of ensuring that your pet will not be upset by the presence of dogs in the reception areas.

Telephone the veterinary hospital beforehand to arrange a visit to meet some of the team. You will probably see more of the receptionist and the veterinary nurses than the vets themselves, so they should be friendly and helpful too. Ask to view the facilities they have. Larger centres are likely to provide more specialist services such as dentistry, ophthalmology, dermatology, internal medicine, cardiology, neurology and behaviour therapy than smaller ones.

When Going to the Vet

Do

- ✓ Take your cat in a carrier or escape-proof box (but with visibility).
- ✓ Be ready to answer questions about your cat's general health and any specific problems.
- ✓ Have your cat's medical records or other relevant documentation with you.
- ✓ Take any samples that the vet requests in a clean, sealed container.

Don't

- ✗ Ask someone to go in your place. You know the most about your cat's symptoms.
- ✗ Call your vet at night or over the weekend unless it's an emergency.

Contacting the Vet

It is always best to make an appointment before going to the vet. This will minimise the time spent waiting. Few vets these days will make house calls except in emergencies or if there are a large number of animals to treat. The vet has all the necessary equipment and drugs in the clinic, making examination and treatment much more efficient. If a house call is absolutely necessary, telephone as early in the day as possible, because it will take up considerably more of the vet's time. If you anticipate that the treatment may take a long time, make this clear when you call to set up the appointment. Have the cat contained in a pen or carrier – often the animal will hide when a stranger arrives.

Take the cat to the vet in an escape-proof box or carrier, and don't send somebody else in your place – you are the only person who can describe the cat's symptoms and make any decisions. Be prepared to give a full history of your cat's complaint and answer the vet's questions as accurately as possible.

Warn the receptionist if your cat is likely to behave aggressively in the consulting room. A frightened cat can be difficult to examine, and a sedative may be necessary. If urine or faeces samples are required, put them in a clean, sealed container. Samples can be collected easily in a clean, dry litter tray, which most cats will use.

At the end of the visit, make sure that you understand the vet's instructions for administering any prescribed medications. Ask the vet or nurse to clarify if you are not sure, and make an appointment for a follow-up visit if required.

▼ *Your vet is an expert who should love animals and want to help. Building a three-way relationship is the key to providing the best care for your cat.*

● ***How often should I take my cat for a checkup?***

An annual checkup combined with booster vaccinations (see pages 86–87) is a reasonable minimum for healthy cats. The young, the old, and cats with existing medical problems should be seen more regularly. This approach should help you to get the most from your vet.

● ***Why does the vet grasp my cat Lucy by the back of the neck even if she is not giving an injection?***

Some cats do not like to be handled by strangers and will attempt to bite or scratch. The loose fold of skin that forms the scruff of the neck can be used to immobilise the patient. This does not hurt the cat.

● ***I have noticed that Tuki, my 3 year-old-Siamese, always leaves damp paw prints on the vet's table. Why is this?***

Cats sweat through the pads when they are stressed – the equivalent of sweaty palms. Alternatively, he may have urinated in his box and got his feet wet.

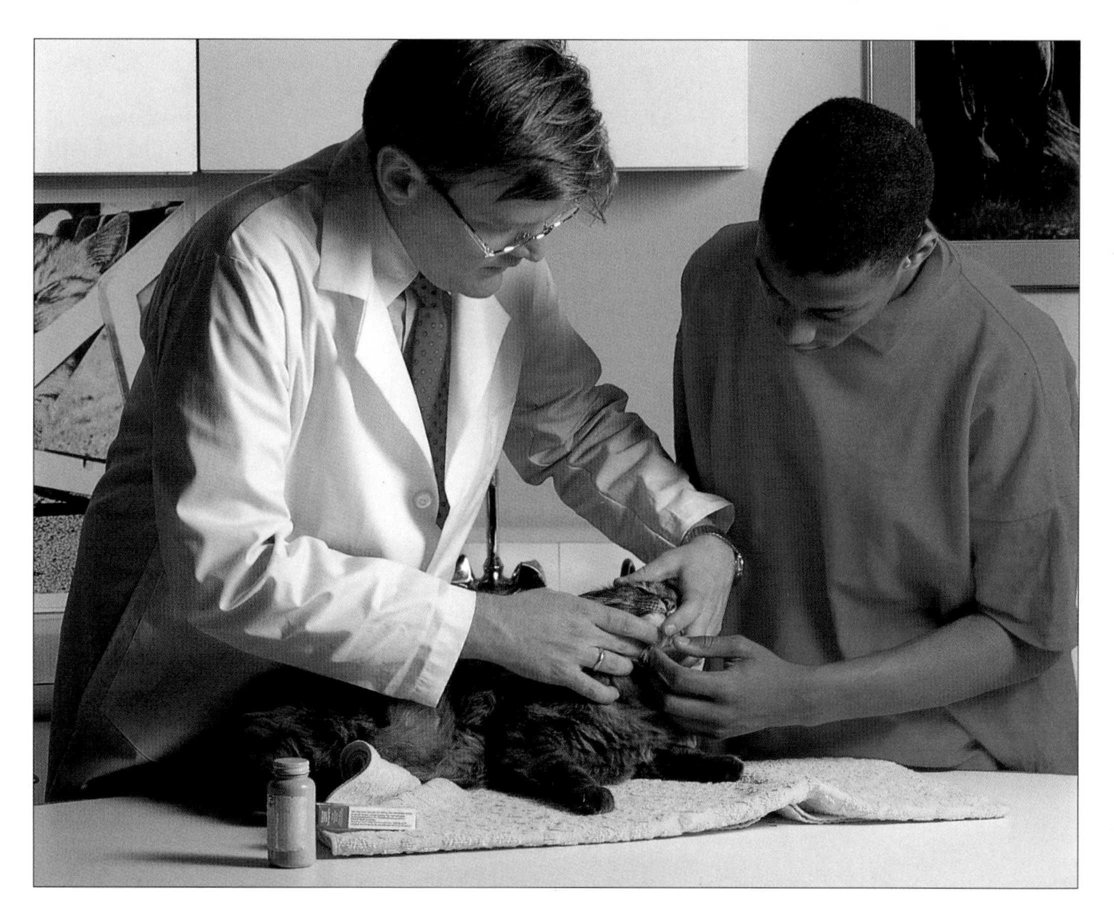

The Anatomy of the Cat

CATS ARE DESIGNED AS HUNTING CARNIVORES, a function for which they are well adapted, and there are remarkably few anatomical differences between your family cat and a lion wandering the Serengeti Plain. Selective breeding for domestic cats has produced some differences – from the long, slim, shorthaired Siamese to the sturdily built American Shorthair and the long, solid, longhaired Persian – but there is much less difference in the size and type of cat breeds than is seen among domestic dogs. Most cats weigh between 2.5 and 5.5kg (6 and 12 lbs).

All cats have an exceptionally flexible skeleton that gives them power, grace and superior athletic ability. The highly flexible spine allows a wide range of movement, and cats actually walk on their toes, effectively lengthening their limbs. In addition, the clavicle (collar bone) is a small piece of cartilage, allowing the scapula (shoulder blade) to move with the rest of the forelimb and so increase stride length. This, together with the cat's powerfully muscular hind limbs and pelvis, allows it to achieve great speed over short distances (up to 48km/30 miles per hour) as well as pounce on its prey in huge leaps. The tail is the cat's 'fifth limb' and assists in maintaining balance as well as signalling the cat's mood.

Cats have little stamina, and most of the time they prefer to conserve their energy; they rarely overexert themselves. Their bones are light; even a slight impact can cause fractures.

Cats' coordination depends on a superb sense of balance, augmented by their nervous system. Cats rarely misjudge a manoeuvre, and if they do, they are usually able to land on their feet.

▼ *Cats are natural athletes and have the ability to arch their backs in both directions in order to sprint and pounce on their prey. They can leap vertically and horizontally and jump up to five times their own height.*

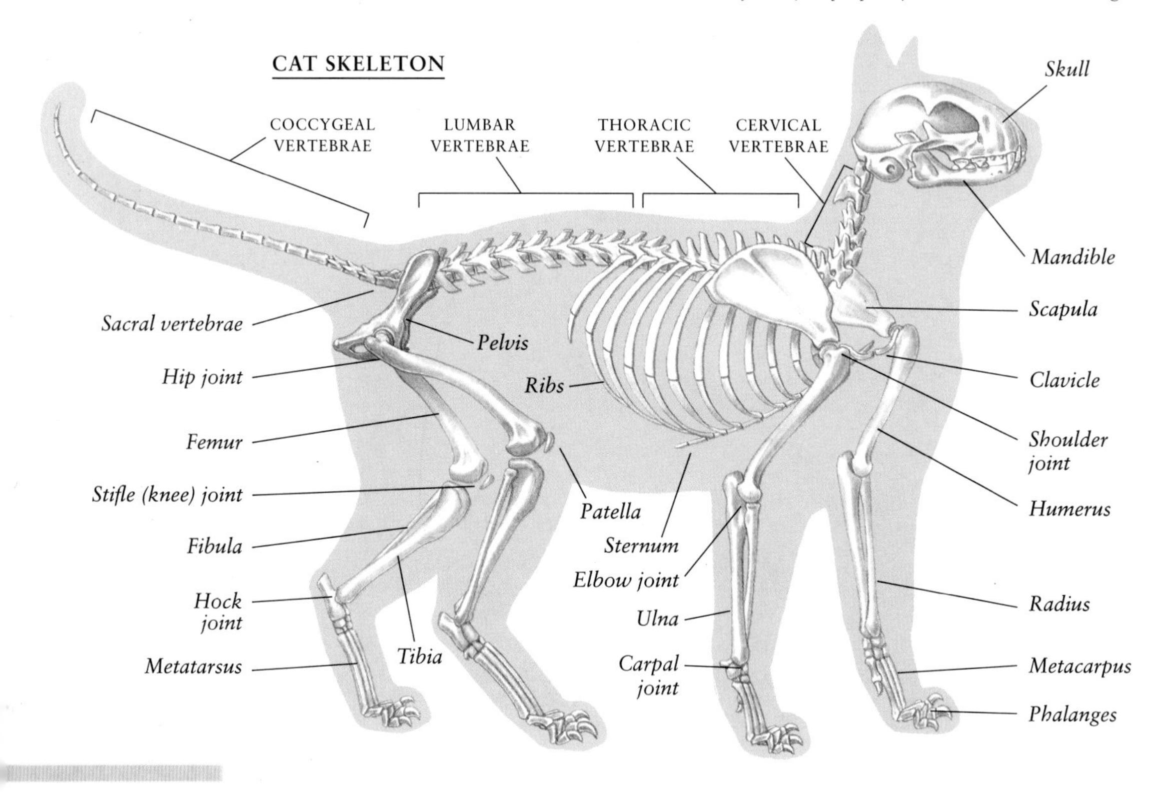

BODY POINTS

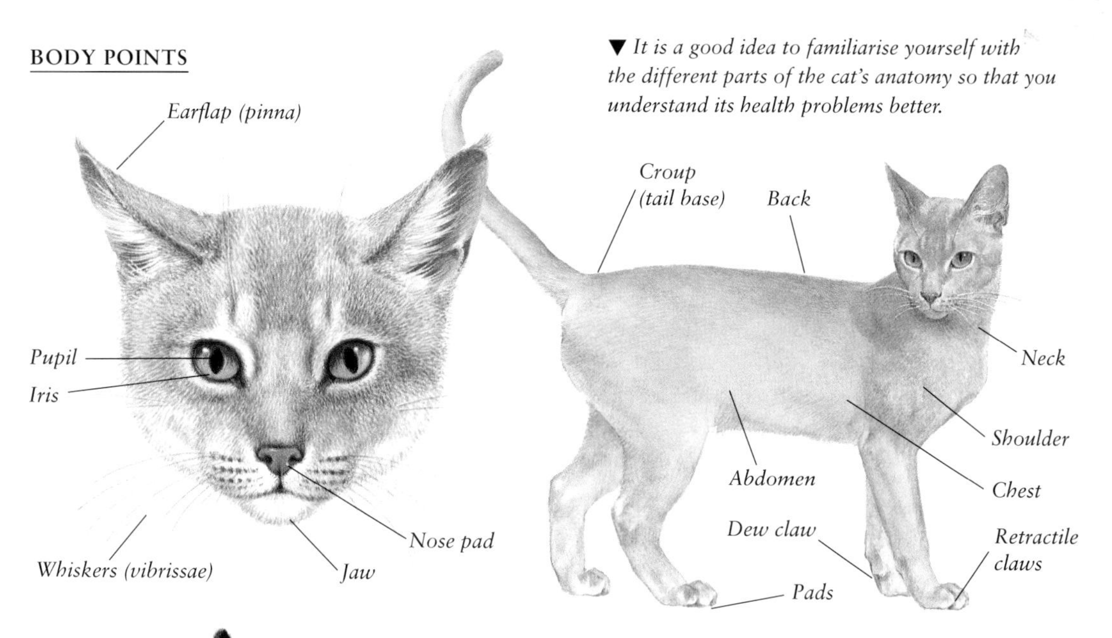

▼ *It is a good idea to familiarise yourself with the different parts of the cat's anatomy so that you understand its health problems better.*

RIGHTING REFLEX

● ***My cat, Tiger, has six toes on each front paw. Is this normal?***

No. Most cats have five toes on the forepaw, including one 'dew claw', and four on each hind paw. However, a large number of cats have extra toes, giving six on each front foot and five on each hind foot – they are 'polydactyl'. This is an advantage in hunting, climbing and self-defence. They may need clipping to prevent overgrowth.

● ***My cat, Marco Polo, fell from a 4th floor window and wasn't hurt; the cat next door fell from the 6th floor and died. How can two floors make that much difference?***

Veterinarians call this phenomenon 'high rise syndrome'. The cat's righting reflex, which allows it to land on its feet, saves it from injury at lower heights, but after five floors it cannot absorb the force of the impact. Curiously, a cat that falls from an even greater height (above 10 floors) quite often escapes serious injury. This is because, once the cat has righted itself, it remains in a fixed position, with its limbs outstretched and head held high. Its muscles relax, and its outstretched limbs help decelerate its fall, lessening the impact when it eventually reaches the ground.

◀ *Cats have a phenomenal sense of balance, and they are nearly always able to right themselves and land safely after a fall. The eyes and balance organs in the inner ear inform the cat where it is in space and how to land on its feet.*

Male Cats

MALE CATS TEND TO BE LARGER AND REACH sexual maturity slightly later than their female littermates, generally at around 6–8 months of age. Some, especially the Oriental breeds, may become sexually active before this. Few physical changes occur at this time, but you may notice that your adolescent male starts to become fuller around the face due to increases in skin thickness and muscle mass in the jaws. This is most pronounced in shorthaired breeds, but may be hardly noticeable in Siamese and Orientals. Another characteristic sign of a male cat's maturity is a thick ruff of fur appearing at the neck.

▼ *Unneutered tomcats can pose many problems for their owners. They are prone to fighting rival males and will mate with any available female. Most people find the odour of their spray unbearable.*

One of the most obvious distinctions between male and female (and between kitten and tomcat) is their behaviour. As he grows into an adult, a male cat becomes more territorial and starts to fight other unneutered neighbourhood males. Though some fights may involve only hissing and spitting, injuries are common. The majority of these will be relatively minor, but fighting can spread fatal viral infections between cats. Treat any injuries without delay (see pages 130–131) and make sure your cat is vaccinated.

The Mark of a Tom

Tomcats are famous for spraying to mark their territory, even indoors. This behaviour is instinctive and guided by hormones; it is not a lapse in house-training. Neutering at about 6 months of age, before the cat reaches full sexual maturity,

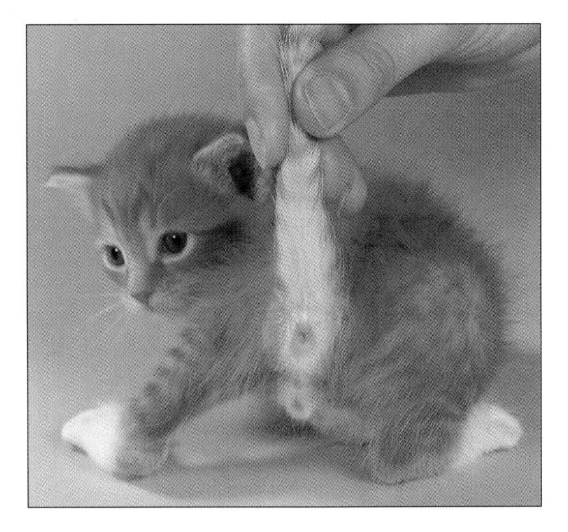

▲ *Sexing a kitten is easily done by lifting its tail. A male can be identified by the small rounded testicles under the anus. The penis is farther down, below the testicles.*

● **Can testicular cancer in cats be treated successfully?**

If detected early enough, the affected testicle can be removed. However, if the cancer has spread to another part of the body, the prognosis is much less hopeful.

● ***I have just adopted a rescued tomcat, Mickey, who has only recently been castrated. He is about 5 years old. Am I likely to have problems with him?***

A lot depends on Mickey's individual character. If he has a placid disposition, he should make a good companion. If he fought and sprayed before you got him, this behaviour will probably continue.

● ***I want to use Prospero, my male Siamese, for breeding, but would like to wait until he is older. How can I control his amorous urges until then?***

You can't. To avoid unwanted pregnancies, you will have to keep Prospero away from any unspayed females by confining him to the house. When his stud days are over, you should have him neutered.

● ***Can two neutered males live happily together?***

Yes, but much depends on whether the cats are related, their individual temperaments, how early they were neutered, and how much space and attention you give them. Littermates are more likely to be compatible. If you wish to keep two unrelated adult cats, neutered males are better than neutered females, but the best combination is a neutered pair.

prevents the production of these hormones and dramatically reduces the incidence of both spraying and fighting, although spraying will sometimes still occur if the cat feels threatened or stressed (see pages 64–65).

Unneutered tomcats tend to wander far and wide in search of females and frequently disappear for days on end. This is always a worry, but especially so if you are the owner of an expensive pedigree male you are keeping as a stud. In this case you will need to take steps to confine your cat at home. If you are not planning to breed your male, neutering helps prevent unwanted kittens in your neighbourhood, makes sure your cat stays close to home, reduces the chances of it getting into fights, and helps prevent some other antisocial habits, such as spraying.

A tomcat that is neutered after the age of 6 months may have already acquired these habits as learned behaviour, and the operation may not change him greatly. For this reason it is best to decide while he is still a kitten whether you will want to breed from him as an adult, and if not, have him neutered as soon as your vet advises. Neutered tomcats usually prove to be highly affectionate and home-loving pets.

Neutering and Health

Castration – the removal of both testicles – is the usual form of reproductive control available for male cats. The surgery is performed under a general anaesthetic and takes only a few minutes. Stitches are often not necessary, and the discomfort to the cat appears to be minimal, though many toms continue to lick the wound for a few hours after. A healthy cat should be back to normal within a day of the operation.

The testicles are situated in the male's abdominal cavity at birth and normally descend into the scrotum as the cat enters adolescence. However, sometimes one or both testicles may remain inside the abdomen, a condition known as cryptorchidism. This makes neutering a more complicated procedure, but it is even more crucial for health reasons: testicles that are retained in the abdomen are more likely to develop tumours than those positioned normally within the scrotum. This condition can be inherited; therefore affected males should not be used for breeding.

Female Cats

MANY PEOPLE PREFER TO OWN FEMALE CATS rather than males to avoid habits like spraying, wandering and fighting (though these problems are much reduced in neutered males). The main disadvantage of owning a female is that, if you allow her to go outdoors at will, she will almost inevitably present you with litters of kittens at regular intervals. If you are not prepared to keep her offspring or find homes for them all, you should have your female cat neutered (spayed); animal shelters are already overflowing with abandoned kittens. A further reason for having your female spayed is that overbreeding shortens the lifespan of the mother.

Unless you intend to use your cat for breeding, neutering should be done when she reaches sexual maturity. This is usually at about 6 months, though it may vary according to how well grown your kitten is. Most cats reach puberty in the early spring, so if your kitten was born late in the previous year, she is likely to be younger at puberty than a kitten that was born earlier. Oriental breeds such as Siamese and Abyssinians sometimes reach sexual maturity as early as 5 months, whereas longhaired breeds such as Persians may not become sexually mature until they are 12 months old, or possibly even older.

Cycles of Activity

In temperate zones of the world, wild and free-living cats are sexually active for about 9 months of the year, stopping during late autumn and early winter, when daylight is minimal. This period of inactivity is less evident in domestic cats, which live at least partly indoors and are exposed to artificial light. Female cats are sexually active for only a short period of each reproductive cycle of two or three weeks. Only during this period, which is known as the 'heat' or oestrus and usually lasts a week, does mating take place. The timing of the cycle is determined by fluctuating levels of sex hormones. In long-haired cats it may take longer than three weeks, while Oriental breeds may have shorter cycles with longer periods when they are in heat. If female cats are neither mated nor neutered, the oestrus periods take place more and more frequently until the cats come to be in heat almost continuously. By contrast, levels of sex hormones in the male cat remain more or less constant, and he is sexually active at all times.

Methods of Preventing Conception

Spaying, which provides permanent contraception, is a routine procedure for removing both the ovaries and the uterus through a small incision. It has the effect of stopping the female's oestrus cycle and makes it impossible for her to have kittens. If you intend to breed from your cat but wish to avoid mating her at a particular time – perhaps because she is still too young, has recently had another litter, or because you are going away – your vet may prescribe hormones to suppress or postpone her coming into oestrus. However, this is only a temporary method. Once she has finished her breeding career, it is best to have her spayed rather than continue to give her hormones on a permanent basis.

Signs of a Female in Heat

A first-time owner who has never seen a female in heat may rush the cat to the vet, believing it is in pain. Heat can begin with little or no warning, but this is what to expect:

- Unusually affectionate behaviour.
- Loud, insistent yowling. This is particularly marked in Siamese and other Orientals, but all females become more vocal when in heat.
- Rolling on her back on the ground or floor in receptive postures with the legs apart.
- Rubbing up against people, surfaces, or even other animals in the household.
- Wandering away if allowed outdoors.
- More frequent urination (to broadcast her scent among the neighbourhood males).

• Are there any risks associated with spaying my cat?

A... As long as the cat is healthy and has no blood-clotting or cardiovascular abnormalities, the risk is low, although there is always some risk with any surgical procedure.The risks of not spaying are greater: a female that is not spayed will be either permanently in heat or repeatedly pregnant, and may develop ovarian cysts or uterine problems.

• *My 13-month-old Chinchilla, Phoebe, has not yet come into heat. Is this normal?*

Longhaired breeds tend to reach puberty late, and 12 months is probably about average. They also show fewer behavioural signs when they are in heat, and these can be missed. Don't worry, but watch for any subtle changes. Puberty can be delayed in small cats with poor growth, and a checkup is a good idea.

• *Do hormone pills for cats have adverse side effects, and if so, how serious are they?*

The progesterone pills commonly used can have serious side effects if they are used in the long term. These include lethargy, obesity, diabetes mellitus and infections of the uterus. Your vet will advise you on how often and for how long your cat should take them.

• *How many litters is too many – when are they a danger to the mother's lifespan?*

A young female can produce three litters a year, but this is too much – it will exhaust her and may affect the kittens' general health. One litter a year for a few years is ideal, but the occasional extra litter a year for a good mother will not cause her too much stress.

• *Do females ever become too old to breed?*

It's best to breed females young. The ideal age is 12 to 15 months for a first litter, and then one litter a year (see above) until the cat is 5 to 6 years old. A female more than 6 years old should not be bred from, but she can still become pregnant; there is no feline menopause. For this reason, a breeding cat should be spayed when her breeding career is over.

• *I've heard that females should be allowed to have one litter before they are spayed. Is this true?*

No. There is no advantage at all to the cat in letting her have just one litter, and there is a considerable disadvantage to you because you will be responsible for the kittens. Of course, if you wish your cat to have a single litter instead of a full-scale breeding career, you can mate her once and then spay her. But wait until she is 12 to 15 months old before mating her.

▼ *Rolling on the ground with her rear end raised gives a clear signal to males that a female is ready to mate.*

Mating

BREEDING NONPEDIGREE CATS IS EASY, BUT BE sure you have homes for the kittens before you allow your female to mate. If she is an outdoor cat and you don't want to her to breed, have her spayed as soon as possible (see pages 76–77).

When mating a pedigree cat, planning is essential to ensure you end up with a litter of saleable pedigree kittens. If you are inexperienced you should look for an expert breeder to work with. It can take a long time to find a suitable stud, so start your search well before your female cat is ready to mate. Before going ahead, always ask the owner for written proof of vaccination status and the absence of the feline leukaemia virus; you will be required to furnish the same. Both cats must be in perfect health.

Once the arrangements for the mating have been agreed upon, observe your female carefully for signs that she is is coming into heat. She will begin to call and roll around a day or two before going into full oestrus. When she begins to press her body to the ground and raise her rear end, she is ready for mating. If you take your female to the stud before she reaches this phase, it may upset her, whereas the urgency of her hormones will overrule her sense of disruption. At first, place her in a pen adjacent to the stud's where they can observe each other. Do not put her with the male until she has had time to settle and has begun to roll and call again.

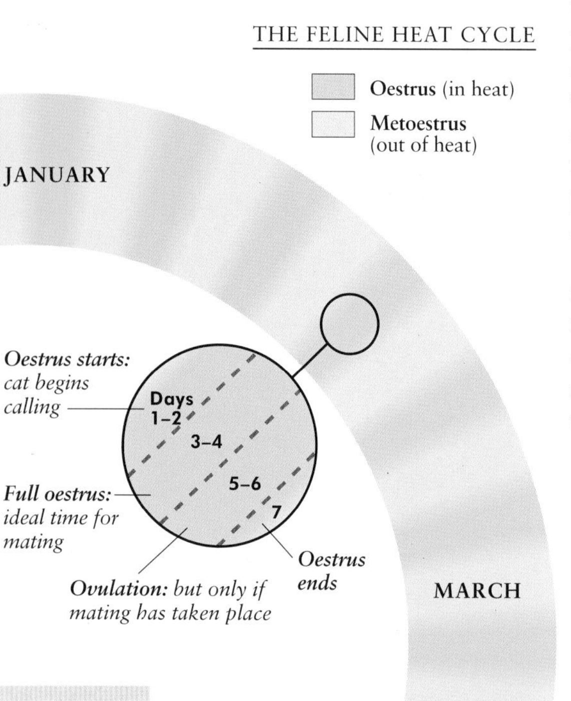

◀ *The female cat's heat cycle (the oestrus and metoestrus) lasts 2–3 weeks. There is a period of hormonal inactivity (anoestrus) in the winter months, September–December in the northern hemisphere. This may not occur when the cat spends much of the time under artificial light.*

A Quick and Risky Business

Although the female appears to play a passive role in mating, initiative and control of the situation are entirely hers. At any time she can turn against the male and attack him. She will reject the first advances as part of the courting ritual, so the male cat, especially if he is experienced, approaches gradually, avoiding eye contact, and grooms the female to relax her. When she is willing, she raises her rear end. Now the male grasps the scruff of her neck with his jaws to prevent her from suddenly attacking him as he mounts her. This bite, together with the stimulation from the male's penis, triggers nerve signals to the pituitary gland in the female's brain, leading to ovulation a day or so later. After a few thrusts the male withdraws quickly and retreats to a safe distance while both cats groom themselves. The female shrieks on withdrawal of the male's spiny penis but will soon begin to roll around again. After a short while the whole ritual is repeated.

Mating can continue like this for three or four days. The female must be kept away from other males until she stops calling and her oestrus is over; she can become pregnant by more than one male and produce a mixed litter, which means that all the careful planning has been in vain.

● *My 18-month-old Persian female, Atossa, has been back to the same stud three times and has not become pregnant. What's the next step I should take?*

There are many contributing factors to unsuccessful matings. Make an appointment with your vet for a checkup to ensure she is perfectly healthy. If all is well and she appears to be having normal heat cycles, make arrangements to visit a different stud.

● *How do I find a suitable stud cat for my female?*

The breeder of your cat may be able to recommend a suitable stud. You could also contact your nearest breed club, who will hold a list of stud cats. You can meet breeders by visiting cat shows – the club should be able to tell you when the next shows are.

● *Can a female cat really become pregnant by more than one male? It sounds impossible.*

It is entirely possible. Female cats ovulate (release eggs for fertilisation) only after mating. The more matings that take place, the more likely the eggs will be fertilised, and fertilisation must occur once for each kitten in the litter. Thus, if more than one stud is used, the kittens are likely to have different fathers.

▼ *The act of mating itself is very brief in cats. After ensuring that the female is receptive, the male mounts, grasping the back of her neck, and penetrates her. All is over in a matter of seconds.*

Mating a Female Cat

- The female should be at least 12 months old, but some longhaired breeds reach maturity later than this. Ask a breeder or vet for advice.
- If she goes into heat before this, ask your vet for a contraceptive (see pages 76–77).
- Choose the breeding premises carefully. You will have to pay a fee for the stud's services, but you will own the kittens.
- Ask to inspect the mating quarters. They should be warm, clean, secure against intrusion and spacious enough for two.
- Females can be very aggressive during mating, so clip her claws beforehand in case she lashes out as the male approaches.
- Mating will need to be repeated several times over a period of two or three days to make sure the female is pregnant, so be prepared to leave her and collect her again when the breeder tells you it is time to do so.
- Keep your female indoors after she returns home until she has stopped calling and rolling.
- If she begins calling again in 2–3 weeks, she has not become pregnant. This is not unusual for first-time matings for males or females; it is best to use an experienced stud if the female has not been mated previously. There is usually no charge for the second mating.

Pregnancy and Birth

▲ *Cats are tender and caring mothers. This concerned queen reaches out gently for her kitten. At 1 day old, it is blind but already recognises her scent.*

ONCE A FEMALE CAT IS MATED, HER HEAT CYCLE usually stops within four days. If fertilisation has occurred, pregnancy begins as the fertilised eggs 'implant' onto the wall of the uterus. There are distinct signs that show your cat is pregnant. At 3–4 weeks her nipples will turn pink or purple and enlarge. Although the fetal cats are tiny, they are surrounded by a tight sphere of fluid that feels like a small, soft marble. This can be detected by your vet – do not prod your cat's abdomen yourself. Soon it will appear swollen, and your cat will gain weight. She may also begin to show maternal behaviour, such as carrying around a toy and trying to 'mother' it.

The average length of pregnancy is about 65 days (9 weeks). Give your cat her usual diet until about 6 weeks. After that she needs to eat more and should be put on a special pregnancy diet, available from your vet. Have her dewormed.

Preparing for the Birth

About a week before the birth, your cat will become restless and start making nests around the house – in wardrobes, drawers, even under the blankets on your bed. Give her several lined boxes to choose as her kittening box and fill the one she selects with lots of paper. As the time for the birth draws near, she will become even more restless and will begin to tear up her bedding.

When the birth is imminent, her restlessness will give way to panting and then to abdominal contractions, at first sporadic and then regular and frequent. The first kitten is often born within 15 minutes of the contractions starting, but sometimes birth is delayed for up to an hour. The mother gives it a quick lick and wash to start it breathing, then continues with her labour.

Unless the mother seems confused, there is usually no need to intervene in the proceedings. Kittens are born in a membrane bag that often ruptures during birth or when the mother licks it. If the mother fails to do this, gently break the bag and rub the kitten with a towel, wiping fluid from its mouth, until it struggles or cries. The placenta (afterbirth) usually emerges with the kitten and may still be attached to it. The mother will normally bite through the cord (and possibly eat the placenta). If she does not do so, use a piece of thread to tie the cord an inch away from the kitten's body and then cut the cord on the placenta side of the knot. Take care not to pull the umbilical cord where it joins the kitten; if you do, it may cause a hernia.

Kittens are usually born at intervals of 15–45 minutes, although queens will sometimes pause for a rest between kittens: after she delivers two or three, there may be a break before the next few emerge. Once all the kittens are born, the mother should recover rapidly and devote herself to cleaning, stimulating and feeding them.

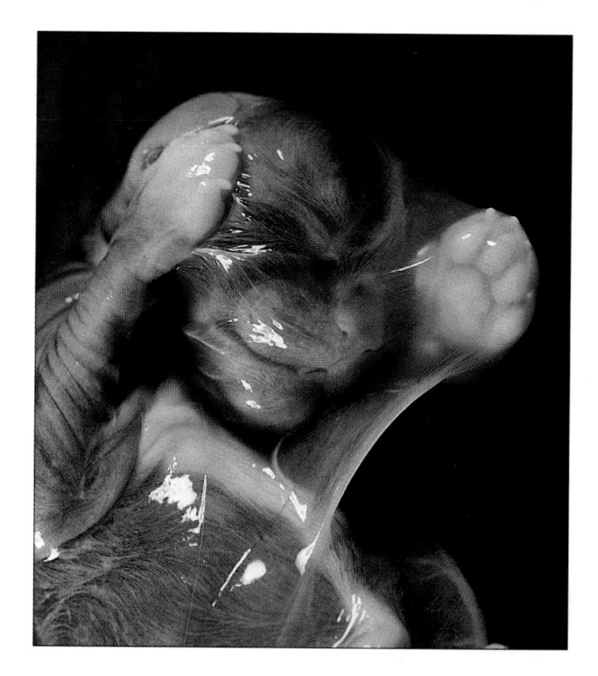

▲ *A tiny kitten just seconds after birth, still within its membrane bag. Every feature can be clearly seen. It will emerge wet and bedraggled from the bag, but vigorous licking by the mother will soon have it moving about and crying, and it will suckle immediately.*

Problems in Pregnancy

- **Miscarriage or abortion:** A bloodstained or dirty discharge at any stage of pregnancy can indicate miscarriage or spontaneous abortion. Abortion is more common in cats infected with feline leukaemia virus (FLV).
- **Reabsorption of fetuses:** Pregnancy will occur only if the queen is able to support the development of the kittens, and sometimes the fertilised eggs stop developing and are absorbed. An ultrasound scan carried out early in such a pregnancy shows more fetuses in the womb than appear on a subsequent scan or at birth.
- **Primary Inertia:** The queen fails to kitten by the 65th day. She should be seen by the vet.
- **Secondary Inertia:** The queen becomes exhausted while delivering a large litter and stops straining. Call the vet.
- **Dystocia (obstruction):** The queen strains unproductively for 30 minutes to an hour without result. Call the vet. Do so earlier if you can see there is a kitten failing to emerge. It may be in the wrong position (see Q&A), or the birth canal may be too narrow for it.

● ***Our Persian queen, Sheba, is due to give birth in two weeks. She seems to have decided on an enclosed box that we have placed in our spare room. It is about 18°C (65°F). Will it be warm enough for her and her kittens?***

No. Kittens are very susceptible to hypothermia, and you may lose them unless you can keep the temperature at 22°C (72°F) at the very lowest. Additional local heat is often needed – you should provide an electric heating pad or, better still, an infrared lamp, positioned just above the kittens to ensure they are warm enough.

● ***Guilia, our Siamese, had her last litter by Caesarian section because one of the kittens was a breech. We'd like her to have more litters. Will it occur again?***

Unlikely. Breech kittens are an accident of nature; they are not due to any flaw in the mother. They occur only rarely, so there is no reason why you should not continue to breed from Guilia.

Raising Kittens

IF YOUR CAT IS FEMALE AND YOU DECIDE TO breed from her, you will have the unique privilege of watching as her kittens develop from tiny newborns. You should have good homes lined up for all the kittens before you allow the mother to become pregnant. All too often owners do not think past the excitement of the birth and the first few weeks, with tragic consequences. If you are not prepared to find homes for all the kittens or keep them yourself, you should have your cat neutered (see pages 76–77).

Early Development

Kittens are born blind and deaf, and rely totally on their mother to anticipate and fulfill all their immediate needs. During the first 3–4 weeks of the kittens' lives the mother cat not only feeds them but keeps them clean. Within a few days they learn to locate her and crawl over to nurse when they are hungry, though their eyes remain closed until they are about 10 days old. By the third week they can stand, hear and see, and by the age of 7 weeks they can run, jump and climb up your curtains.

Between the kittens' birth and the time that they are fully weaned, their mother should be supplied with as much of a high-quality growth-formula food as she wants to eat. Alternatively, the mother cat may be fed a commercial kitten diet while nursing. Nutrition is essential for her well-being and the kittens'. Fresh water to drink must always be available.

► *A tabby mother cat suckles her 5-week-old kittens, who rely on her entirely for their needs. At birth they quickly learn to find her to nurse and will continue suckling until they are 8 weeks old.*

Kittens' eyes start to open at about 10 days, but they cannot focus on objects until they are 4 weeks old. Once they can see properly, the kittens will become more adventurous and begin to play. At this stage you may begin to offer them solid food (see chart on page 84). This may be easier if the mother is already eating kitten food. Once the kittens have begun to eat solids, you can introduce them to the litter box by placing them on it after each meal. They will eat a little more solid food each day but will continue to nurse for several more weeks. Gentle encouragement is beneficial, but do not force them. They will eat more solid food when they are ready.

Kittens grow quickly from the age of 5 weeks. It is important to provide a variety of foods, but the biggest part of their diet should consist of a properly balanced growth food for kittens. The stomach of a young kitten is very small, only about the size of a walnut, so kittens must be fed small meals at frequent intervals. Overfeeding will cause stomach upsets.

Hand Rearing Kittens

If a mother cat is unable to rear her kittens for any reason, it may be necessary for you to rear them by hand. This is rewarding, but it demands a total commitment for the first few weeks. They must be kept at an average temperature of 24°C (75°F) with an overhead infrared heater or another suitable heat source. A hot-water bottle wrapped in a soft, woolly cover makes a good 'surrogate mother' for them to snuggle against.

The kittens must be fed from a syringe or special feeding bottle every 2 hours day and night. Give each kitten approximately 1 tsp (5mls) of a high-quality substitute milk, gradually increasing the amount. Be careful not to let the kittens drink it too quickly; if they do, they may inhale the milk into their lungs and choke on it.

▲ *Hand rearing kittens can be a very rewarding experience. This will be necessary if the mother cat is unable to rear them herself. Feeding is from a specially adapted bottle or syringe every 2 hours.*

● ***Our cat gave birth in our bedroom, but now she moves the litter to a new 'nest' every day. Is something upsetting her?***

This is a relic of instinctive behaviour aimed at keeping the kittens safe from predators. Move the original kittening box to whatever new site she chooses.

● ***My 6-week-old kitten keeps trying to eat the cat litter. What can I do to stop her?***

Kittens being weaned often go through a stage of 'testing' different substances, including litter. Gently discourage your kitten by removing any litter found in her mouth and moving her away, unless you have just fed her and are litter training her; in that case, put her back on the litter. Hygiene is very important. Clean the litter as soon as it is soiled and use a safe, nontoxic litter. Finally, if you have not already dewormed her, ask your vet to recommend a safe deworming product.

● ***How long do I have to wait before I allow my newborn kittens to go outside?***

Kittens should not be allowed outside until 7–10 days after their vaccinations (see pages 18–19). Vaccination is normally done at 9 and 12 weeks, so your kitten must stay in until it is at least 12 weeks old.

After they are fed each time, the kittens must be encouraged to urinate and defecate by imitating the effect of their mother's tongue. Use damp cotton wool soaked in warm water to gently stimulate the abdomen and under the tail.

It is helpful to give the kittens at least one meal of their mother's milk, if she is still nearby, by squeezing milk from her teats and feeding it to the kittens. If the kittens have not been able to feed from the mother cat at all, they will not have had any of her colostrum, which is the first milk that normally gives vital natural immunity against certain bacteria. Without this protection the kittens may be more susceptible to infection. In this case, be especially careful to keep all utensils sterilised, particularly feeding bottles and nipples. It is important that hand-reared kittens be vaccinated as early as possible, usually at 8 weeks. You should contact your vet in advance to make an appointment.

The Importance of Early Handling

The most important phase for handling is from 2–7 weeks, when the kitten is usually still with the breeder. Once you bring it home, the kitten's ease with people will be greatly enhanced if it is frequently handled by others as it grows.

Kittens should be gently handled several times daily for a few minutes to ensure that they grow up friendly and tolerant of the contact necessary for health care and grooming. This handling can include gentle restraint, holding the paws, touching the sensitive underbelly and flanks, as well as gentle grooming. This is particularly important in longhaired kittens.

▼ *A mother carries her 2-week-old kitten by grasping the scruff of its neck with her jaws, causing it to go limp and relaxed. An adult cat will duplicate this reflex if you pick it up by its scruff.*

At six weeks kittens are scampering around, playing with each other and everything else that looks like fun. Introduce toys at this stage but ensure that they are safe. Use disinfectant to keep everything clean. Some disinfectants may be poisonous to cats, so ask your vet if you are not sure what to use. Kittens at this age will sniff and lick everything they encounter, including the litter in

Weaning a Kitten

Age	Meals per day	Food
3 weeks	4–6 from a saucer	Tinned or powdered cat's milk formula, diluted with water
4 weeks	As above	Powdered cat's milk mixed with baby cereal or baby food (pureed fish, meat, or cheese), tinned kitten food, or finely chopped cooked meat
5 weeks	4–5	Finely chopped cooked meat or complete kitten food recommended by your vet
6–8 weeks	3–4	Increase quantity of food (as above) and restrict access to nursing
After 8 weeks	As above	Nursing should stop. Milk is no longer necessary if the kitten is being fed a balanced formulated diet for kittens

● ***My kittens have been hand reared. Do I need to wean them any differently?***

Weaning for hand-reared kittens should begin at 3 weeks, but begin by adding a half-teaspoon of fine-textured baby cereal or pureed baby food to the bottle. After a few days you can follow the normal routine (see chart).

● ***Our kittens are now 6 weeks old. Three of them are eating solid food, but one does not seem interested. Could there be something wrong with him?***

At this age some kittens still prefer their mother's milk to solid food. Spend a little extra time with him and tempt him with tasty morsels. If you are still concerned, let your vet check him over to make sure all is well.

● ***If prospective owners are coming to inspect my kittens, when should they be allowed to come?***

The ideal time is when your kittens are exploring and their personalities are emerging. This is at 7–10 weeks. Pedigree kittens are not usually released to new owners until they are 12 weeks old, after they have had their second vaccinations.

the litter box. They are hyperactive and constantly explore every nook and cranny of your house. Now is the best time to train your kitten not to sharpen its claws on your furniture, climb your curtains, or jump onto shelves full of precious objects (see Preventing Behaviour Problems, pages 44–45). Misbehaviour that seems cute now may become an annoyance if it persists.

Playing games will also be beneficial to your relationship with your cat. If a kitten becomes overexcited or aggressive, stop and walk away. If it plays gently, however, make sure you praise it in order to reinforce this good behaviour.

Rewards such as food treats should be given for desirable behaviour such as coming to you when you call, while a sharp 'No!', or a thump on the tabletop, is usually sufficient to stop bad behaviour such as climbing on kitchen counters, scavenging food, or scratching the furniture.

▼ *Nothing could be cuter than these two kittens romping under a blanket. By this age they are investigating everything, so make sure your prized possessions are out of reach.*

Infectious Diseases

THERE ARE A NUMBER OF INFECTIOUS DISEASES that affect cats. Some of them are fatal. It is possible to vaccinate your kitten against a number of these as soon as it is old enough. Thereafter, a yearly booster will guarantee lifelong protection.

Feline Influenza (cat flu) is the most common virus in cats, causing sneezing, fever, and a runny nose and eyes. Depending on age and general condition, the cat's immune response will normally fight off the virus, but antibiotics may help prevent secondary bacterial infections. Your cat can be protected by vaccination.

Feline Enteritis (enteric corona virus) is less common than flu but can kill kittens and young cats in a few hours. The virus is spread in faeces, and the illness begins with severe vomiting and diarrhoea. The cat rapidly becomes dehydrated, and may collapse and even die. Treatment is with intravenous fluids. A vaccine is available.

Feline Leukaemia, the most deadly infectious disease of cats, is usually caused by a virus that destroys the cat's immune system. The virus is spread in saliva. Symptoms include rapid weight loss or disease of the mouth and gums. There is no cure, but supportive treatment in the early stages can prolong life. A vaccine is available.

Feline Chlamydia causes conjunctivitis, a painful inflammation of the eye that is highly infectious. Your cat can be protected by vaccination.

Feline Immunodeficiency Virus (FIV), similar to HIV in humans, is transmitted through bites or sharing feeding bowls, and is passed from mothers to their litters but not thought to be sexually transmitted. After a period of mild illness and fever, with enlargement of the lymph nodes, severe immunosuppression follows – the cat will not eat, loses weight and has gum infections, diarrhoea, vomiting and skin or eye problems. Antibiotics or corticosteroids may temporarily reduce the severity of these symptoms, but there is no vaccine and FIV is invariably fatal. Known carriers of FIV (who may be healthy) should be kept permanently away from uninfected cats.

▲ *This kitten has feline influenza (cat flu) with telltale watery eyes. Cat flu is highly contagious and can be spread in pet shops or kennels, at shows, or even at the vet's. Spare your cat discomfort by having it vaccinated.*

What You Should Do

- ✓ Arrange to have your kitten or cat vaccinated as soon as you acquire it. Do not let it outside until it has completed the full course of shots.
- ✓ Do not let an unvaccinated cat have contact with other unvaccinated cats.
- ✓ Take your cat for booster shots every year.
- ✓ Neuter your male. Intact toms fight more, and bites are often the point of contact that spreads disease.
- ✓ Wear disposable gloves to change cat litter. Discard them afterward and wash your hands.
- ✓ Change the cat litter at least once a day. Pregnant women should find someone else to take responsibility for this chore.

● *My kitten is due to have its first vaccinations next week. Are there likely to be any side effects?*

Some kittens and cats may appear sleepy and under the weather for a short time after vaccination. If this lasts more than a day or two, you should contact your vet. Cats may infrequently develop immune-mediated diseases or vaccine-associated sarcomas, but these risks are very slight compared with the far greater benefits of complete protection.

● *Do the vaccines ever fail to work?*

Occasionally, either because the cat's immune system is damaged or a kitten's maternal immunity is too strong for the first round (see pages 18–19). The second vaccination is usually successful. Vaccines may become ineffective if they are improperly stored, transported, or administered.

● *What happens if I forget the yearly booster?*

Your cat's immunity will disappear, and it will probably be necessary to restart the vaccination programme from scratch with two injections, as for kittens. Your vet should normally send you a reminder each year when your cat's vaccinations become due.

● *Are there homeopathic alternatives to vaccines?*

There is no scientific evidence that homeopathic alternatives to conventional vaccination actually work, and all such claims should be treated with caution.

▲ *Chlamydia in this young kitten has caused conjunctivitis and flulike symptoms. It can be treated with antibiotics.*

Feline Infectious Anaemia is a disease of the red blood cells caused by the Hemobartonella bacteria. It is thought to be transmitted from infected cats by fleas, but may also be passed on to kittens from the mother via the placenta. Symptoms are anaemia, lethargy and weight loss. There is no vaccine but antibiotics are usually effective. Recovered cats may be carriers.

Feline Infectious Peritonitis (FIP) is caused by a coronavirus transmitted by licking infected faeces or inhaling virus-laden droplets. The disease is called 'the great impersonator' because any system may be affected, and in its early stages it can present generalised symptoms of slight fever, dullness, loss of appetite, diarrhoea and mild chest infections. Later on fluid may accumulate in the abdomen, chest, or around the heart. Problems of the eyes and central nervous system may also occur, or may be the only sign. Vaccines are ineffective. Treatment with corticosteroids and antibiotics will help in the short term, but this disease is ultimately fatal.

Toxoplasmosis is spread by a parasite found worldwide in all warm-blooded animals (including humans) and carried in faeces. About 50 per cent of cats have toxoplasmosis at some time in their lives, but it causes symptoms primarily in cats with damaged immune systems; diarrhoea is the most common symptom. Once shed in a cat stool, the organism must remain in the faeces for 24 hours before a human can be infected, so if litter is disposed of quickly, there is no risk to human health. However, pregnant women should not handle cat litter, because toxoplasmosis can infect the unborn fetus. There is no available vaccine. Treatment with an antiparasitic drug will usually cure the problem in otherwise healthy cats.

Rabies, a virus that affects the central nervous system, is spread by the bite of an infected animal and can be passed to humans. It can take from 10 days to 6 months for symptoms to appear: the cat drools at the mouth, and its behaviour changes; some cats become demented, others withdrawn. Death usually follows within 5–7 days. Annual vaccination is essential unless you live in a rabies-free country protected by strict quarantine laws, such as the United Kingdom, Japan, New Zealand, or Australia.

Common Parasites

CATS ARE HOSTS TO MANY KINDS OF PARASITES. Some are insects that live on or close to the skin and feed on the animal's blood; others are parasitic worms that live in the gut and other parts of the body. Some of the parasites that live on or in cats can, in certain conditions, infect humans and can be a potential hazard to your family's health. Make prevention a routine part of caring for your cat. Do not wait until your pet is showing signs of infestation before taking action.

Parasite populations are influenced by variable factors such as the temperature, climate and the density of the local population of cats. Fleas, for example, need warm temperatures to reproduce (though central heating means they are becoming a year-round problem even in cool temperate zones). Roundworms are more of a nuisance in urban areas than in the country. Your vet is the best person to advise you on protecting your cat and scheduling preventive programmes.

Outside the Body

Fleas are a very common parasite of the cat. Cats often become allergic to their bites, and they are responsible for a large number of skin problems (see pages 92–95). The females lay between 20 and 50 eggs a day, which fall off the cat into the

● ***What's the best way of checking if my cat has fleas?***

... Place the cat on a white surface and brush its coat carefully so that any small black dots falling from its coat will show up clearly. Use a damp sponge or cotton wool to pick them up. If they smear and turn reddish brown it is a sure sign they are specks of flea dirt, which consist of partially digested blood. You can sometimes see the flea dirt as black dots in the cat's fur, especially if it has a white coat. They are much easier to spot than the fleas themselves.

● ***How do I deal with a tick on my cat's skin, should I find one?***

Do not attempt to remove the tick while it is alive. The easiest way to kill it is with rubbing alcohol or flea and tick spray. Carefully pull the dead tick out with fine tweezers or use a tick-removing tool, available from your vet – you slip it between the tick and the skin and then rotate. Do not squeeze the tick as this will release more poison into the cat.

● ***How effective are flea collars?***

On the whole, they do not work very well and certainly do not eliminate the need for routine spraying and vacuuming of the home environment. The medication used in the collars can cause allergies and skin irritation in some cats.

External Parasites

PARASITE	LIFE CYCLE & SYMPTOMS	PREVENTION & TREATMENT
Fleas	Adult fleas find cat host to feed on. Bites can cause severe allergy, with skin scaling and hair loss	Give regular antiflea protection. Check cat regularly for flea dirt. Vacuum and wash bedding. Spray carpets
Lice	Spend whole life cycle in cat's hair. Irritation and itching. Can lead to anaemia in severe cases	Killed by most antiflea treatments
Ticks	Attach themselves to skin and drop off after feeding. Localised irritation. Saliva can transmit other diseases, e.g., Lyme disease	If living in tick-infested area, use an anti-tick wash or topical preparation on outdoor cat. Check coat daily. Kill and remove any ticks
Mites	Demodectic and sarcoptic mites burrow into skin, causing irritation and skin scaling. Ear mites cause scratching and head shaking	Apply mite-killing wash or eardrops, as directed by vet

environment (the cat's bedding, your carpets and furniture). In warm weather the eggs will be hatched in 2–12 days. After passing through the larval and pupal stages, the adult flea emerges and immediately seeks a host to live on. The entire life cycle may take as little as 3 weeks to complete, but the pupae are capable of remaining dormant for as long as a year in cool weather. To make sure of controlling fleas effectively, all animals in the house must be treated regularly with an antiflea product. These act by killing fleas or preventing them from breeding by making their eggs sterile. They are available in various forms: as injections, sprays, topical lotions, and as capsules that are opened onto the cat's food once a month. If your cat becomes infested with fleas, you must spray bedding, curtains, rugs, carpets and furniture to destroy all the eggs and larvae in the cat's environment.

Lice, small, pale-brown insects that move slowly up and down the cat's hairs and on the skin, are much less common than fleas. They spend their entire life on the cat, laying their eggs on its fur. Antiflea treatments usually kill lice too.

Ticks are tiny creatures related to spiders. They live on vegetation and get caught up in the cat's coat as it brushes past, attaching themselves to its skin. As they feed, their bodies swell up with blood and they look a little like small grey peas. They will drop off after a few days but should be removed at once if seen on your cat. Most ticks cause localised irritation, but some can affect their victims more severely, causing paralysis or debilitating illness. For instance, in the US, the deer tick carries Lyme disease, a slow-developing illness characterised by a skin rash and joint inflammation. If you live in a tick-infested area, search your cat's coat daily and use a protective wash, spray, or spot-on treatment regularly.

Mites, which are barely visible to the naked eye, are passed on by direct contact between animals. They feed on the skin or the lining of the ear canal, causing scaling and irritation (see pages 92– 95). They can be treated with washes, sprays, or eardrops.

THE LIFE CYCLE OF THE FLEA

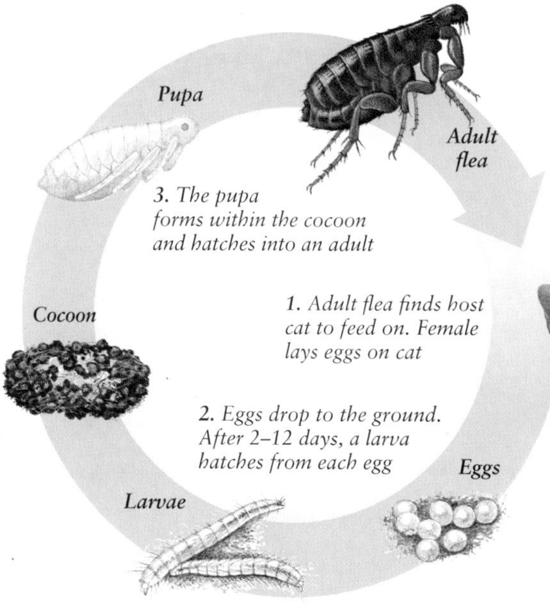

Adult flea

1. *Adult flea finds host cat to feed on. Female lays eggs on cat*

Eggs

2. *Eggs drop to the ground. After 2–12 days, a larva hatches from each egg*

Larvae

Cocoon

Pupa

3. *The pupa forms within the cocoon and hatches into an adult*

▲ *Most of the flea's life cycle is spent in the environment. Once hatched, the adult flea spends its entire life on the cat, feeding on its blood.*

▶ *A cat is treated with a pump-action flea spray; many cats find the hissing sound of an aerosol alarming. Treat the area around the face and ears by spraying onto a cloth and wiping over the cat's fur.*

Inside the Body

Parasitic worms are picked up by cats from the environment. Outdoor cats, in particular, are liable to ingest them by eating infected prey such as small rodents, birds and beetles. The worms live within the cat's gut and other parts of the body and can cause debilitating symptoms such as abdominal pain, diarrhoea, anaemia and weight loss. It is important to protect your cat adequately against them. The cat passes the eggs, and sometimes the worms themselves, out of its body in its faeces. To prevent spreading parasites, always dispose of cat faeces by flushing them down the toilet.

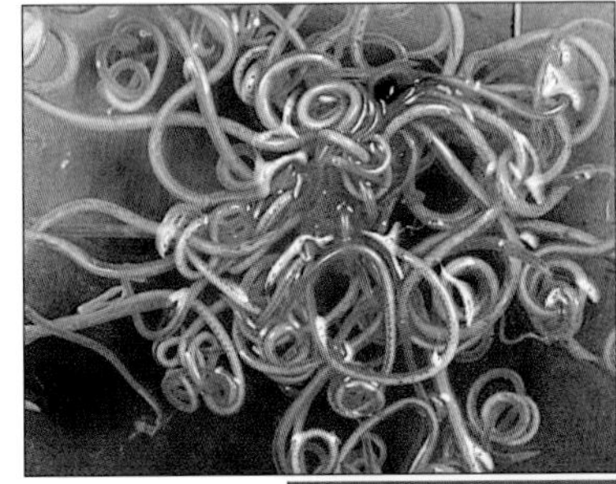

◀ *A mass of roundworms, which infest most kittens. A deworming tablet is given through a plastic syringe, or pill popper (below).*

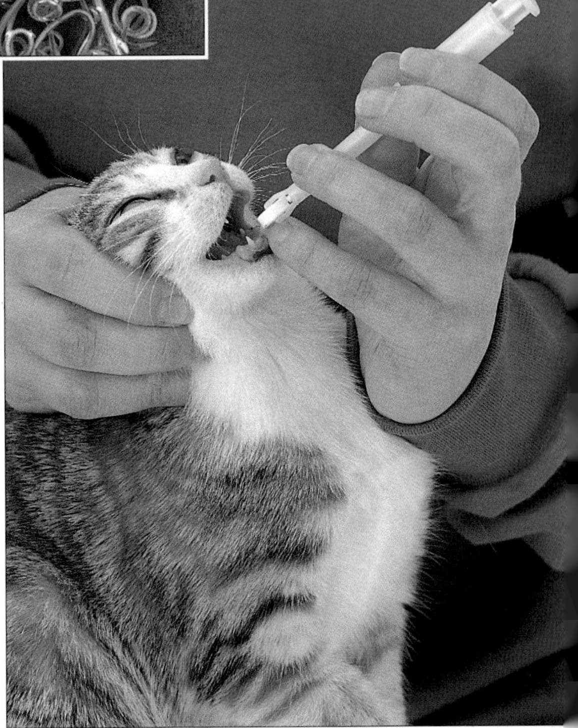

Roundworms are round and white, often coiled, and up to 15cm (6in) long. They are readily seen in the cat's faeces or are occasionally vomited up. Two types are found in cats, Toxocara cati, the most common and (much more seldom) Toxascaris leonina. Because the larvae are present in the queen's milk, almost all kittens are infected with T. cati by the time they are 3–4 weeks old. Heavily infected kittens are usually debilitated, have diarrhoea and often appear potbellied. All kittens should be dewormed at regular intervals from 3 weeks to 6 months of age. Thereafter, they should be tested and/or dewormed every 6 months (outdoor cats) or 12 months (indoor cats): ask your vet's advice. Queens should be treated during pregnancy and lactation.

Tapeworms have a flattened, segmented body that grows from a small pointed head embedded in the intestinal wall. Dipylidium, common in cats, can reach a length of 50cm (20in). As the tapeworm grows, segments break off the from the rear and are passed through the anus. The segments are mobile and full of eggs but soon dry – when they look like grains of rice – and then burst, releasing the eggs into the environment. The eggs are eaten by flea larvae, where they develop into cysts. They remain dormant until an adult flea is eaten by a cat during grooming, whereupon the cyst develops into an adult tapeworm in its intestine. The flea is the only intermediate host – there is no cat-to-cat spread.

Taenia, a longer and thicker tapeworm than Dipylidium, is picked up from infected prey such as small rodents and carrion. If your cat is a regular hunter, you must make sure it is routinely dewormed. Heartworm preventive treatments also protect against tapeworm. Avoid feeding your cat offal or raw meat.

Hookworms and whipworms are intestinal worms found in parts of the US and Australia, picked up from eggs licked from the ground or grass. Larvae of the Ancylostoma hookworm can burrow their way through the skin, especially the feet, causing severe irritation. Kittens are infected through the milk of their mothers. If you live in an infected area, it is essential that your cat is given full protection.

Heartworms – small parasitic worms that live in the heart – less commonly infect cats than dogs. The cat becomes infected with the heartworm larvae through the bite of an infected mosquito.

They enter the circulation and develop to maturity in the heart, where they obstruct the blood flow through the lungs and eventually cause heart failure. Symptoms include breathing difficulties, weight loss and a buildup of fluid in the abdomen. Prevention, in the form of a daily or monthly tablet, is essential in areas where heartworm is endemic, including most of the US and parts of Canada.

Lungworms live in the lungs of the cat; in large concentrations, they cause respiratory disease and coughing. The larvae are passed out in the cat's faeces and are eaten by a slug or snail, which in turn is swallowed by a bird or rodent. The cat then ingests the larvae in hunted prey, and the cycle is completed. A specific deworming treatment is needed from your vet.

Toxoplasma is a minute parasite, called a protozoan, that lives in the intestines of the cat. It uses many mammals, including humans, as its intermediate host. A cat becomes infected by eating infected prey species or uncooked infected meat, or by ingesting oocysts present in other cats' faeces. Diarrhoea is common in acute infection. To avoid infection, use disposable gloves when handling your cat's litter box. Dispose of the faeces daily, as toxoplasma cysts passed by the cat are not infective for 24 hours. Pregnant women are at particular risk because toxoplasma can cause abnormalities in the unborn foetus.

● ***How do females pass on roundworms to their kittens?***

Roundworm larvae can remain dormant in body tissue. When a female becomes pregnant, hormone changes awaken the dormant larvae. Some larvae present in the mammary tissue are passed on to the kitten after birth in the mother's milk.

● ***I took Pericles, my tom, to the vet because he was scratching his bottom. The vet did a faecal test to see if he has tapeworm. Meanwhile, he said, I should improve Pericles's antiflea treatment. What was the reason for this?***

The shedding of the tapeworm segments can sometimes cause irritation around the anus, and frequent washing of the area is often one of the first signs that a cat has tapeworms. As the eggs are passed on by flea larvae, it is essential to keep the cat's environment free of fleas to avoid reinfection. As well as treating Pericles, you must wash and spray his bedding and vacuum the house thoroughly to make sure all flea eggs and larvae are destroyed.

● ***I can never get Godzilla to swallow his deworming tablets. Have you any suggestions?***

A pill popper that deposits the tablet at the back of the throat sometimes helps, or you can try crushing the tablet and mixing it with food. Deworming medicines also come in the form of a paste that you add to the cat's food. Treatment for tapeworm can be given in the form of injections.

Internal Parasites

PARASITE	SITE & SYMPTOMS OF INFESTATION	PREVENTION
Roundworm	Intestine and lungs. Inflammation of stomach; diarrhoea; pain	Regular deworming from 14 days old. Stool (faecal) tests at annual checkup
Tapeworm	Intestine. Diarrhoea; vomiting; weight loss	Good antiflea control and regular deworming. Avoid raw meat and offal.
Hookworm	Intestine. Inflammation; diarrhoea; blood in faeces; weight loss; anaemia	As part of regular worm prevention. Stool test at annual checkup
Whipworm	Intestine. Diarrhoea; blood in faeces; poor coat; weight loss; anaemia	As part of regular worm prevention. Stool test at annual checkup
Heartworm	Heart and lungs. Cough; weight loss; weakness; abdominal swelling; anaemia	Daily/monthly tablet. Blood test at annual checkup. Start prevention 2 weeks before visiting an endemic area; continue for 90 days after
Lungworm	Lungs. Coughing	Specific deworming treatment. Transtracheal wash or special stool sample often required
Toxoplasma	Intestine. Often symptomless. Can cause heart and liver disease, pneumonia	Take precautions when handling litter trays and faeces

Skin and Coat Problems

Cats are prone to several skin conditions that cause itchiness, but they are not always easy to detect because cats are fastidious about their coats and will naturally spend a lot of time licking, nibbling and grooming their fur. You may be alerted to an incipient problem if your cat seems to be grooming itself more than normal and frequently regurgitates fur balls. Other signs are scratching, rubbing, or chewing at its own skin. However, some cats seem to scratch themselves only in private so the problem does not become immediately obvious. It is not until bald patches, scabs, scaling, or areas of inflamed skin appear, singly or in combination, that a visit is made to the vet.

Identifying the precise cause of a skin problem is not always easy as the symptoms rarely point to a specific disease. The vet will inspect the coat for broken, firmly attached stubby hairs, and often examine the hairs under a microscope for signs of fracture: this will indicate if there is itching and scratching present. The vet will take several other factors into account when reaching a diagnosis: is the itchiness confined to one part of the body only, such as the ears or feet, or is the whole body affected? Does there seem to be a seasonal link? Are other pets or humans in the household also showing symptoms of itchiness? Have there been any recent changes in the cat's diet or its environment, such as a new brand of household cleaner? Has the cat been in contact with any chemicals?

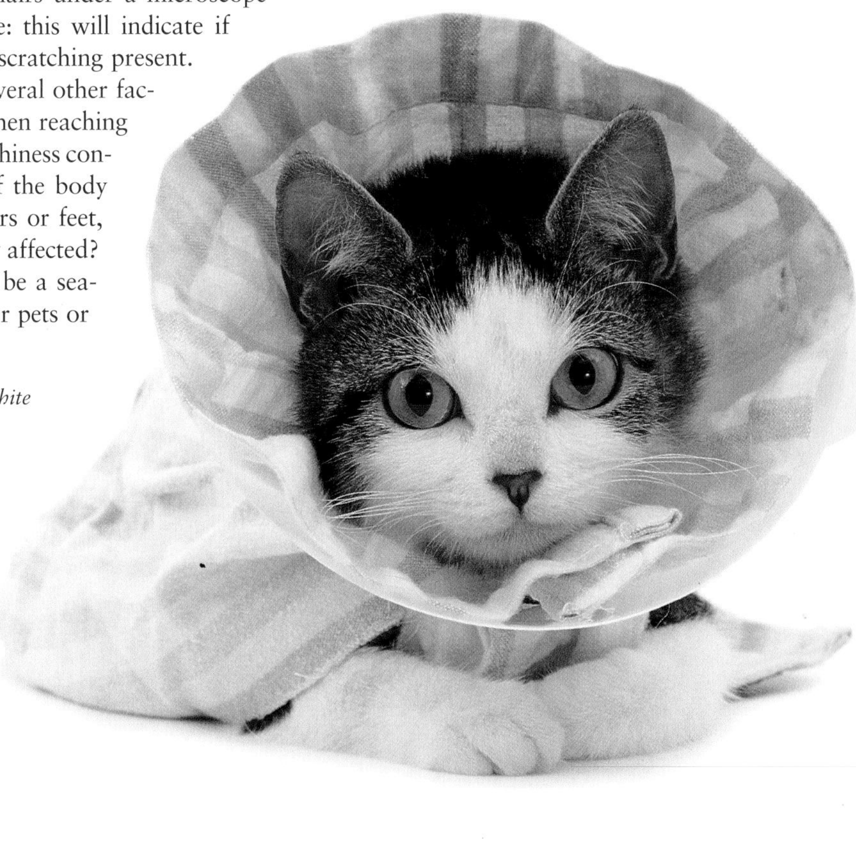

▶ *A small tabby-and-white with multiple allergy-induced eczema wears a 'pajama' bag and Elizabethan collar to protect the skin and prevent licking during treatment.*

Fleas and Mites

In most cases the vet will carry out tests such as skin scrapes and coat brushing to establish if parasites are present. Fleas (see pages 88–89) are a common cause of feline itchiness. All cats are irritated by fleabites, but some individuals may develop a hypersensitivity to the flea saliva in the bites. Even a single bite can cause intense itching, with severe scratching and self-mutilation. The coat becomes dull, and large scaly, spotty areas are visible on the skin. When dealing with cases of flea infestation, be sure to treat all in-contact animals to kill the adult fleas. The cat's bedding

● *Selima, my cat, has an allergy to house-dust mites. How can I help her?*

Wash her bedding weekly, at a high temperature. Don't let her creep into a heated linen cupboard to sleep, as the mites flourish in the warm atmosphere. Keep her out of the room when you vacuum (equipped with an anti-allergy vacuum if possible) and use a damp cloth for dusting. Your vet may prescribe 'allergy shots' to desensitise her, or corticosteroids and antihistamines.

● *Our tom, Tumbleweed, loves hunting in nearby fields. This summer he started itching and worrying at his ears and face, and when we looked closely, we could see lots of tiny orange specks moving about. What were they?*

It sounds as if Tumbleweed has been attacked by harvest mite larvae, often known as 'chiggers'. In late summer they attach themselves to the skin of animals, often cats and dogs, causing intense localised itching. Tumbleweed probably picked them up on his face and ears when stalking through long grass.

● *I've recently noticed Tigger, my cat, dragging his bottom across the carpet. Why is he doing this?*

This may be a sign that the skin around the anal area is irritated. There could be a number of causes, such as flea or worm infestation, or diarrhoea, but this behaviour is a typical sign of overfull anal sacs, the scent glands on either side of the anus. Sometimes they become swollen and infected, and will need emptying by the vet. If they become chronically impacted, you may wish to consider having the sacs surgically removed. Cats can easily live without them.

and your rugs and curtains should be sprayed so that the dormant eggs and larvae are destroyed. Check human family members, too – the fleas will sometimes attach themselves if no other animal hosts are available.

Fur mites (Cheyletiella) are another cause of skin irritation in cats. They are highly contagious and will also bite humans, so if any member of your family has a skin rash, be sure to mention this to the vet. The cat often has a rash and scaly skin (dandruff), especially around the neck and along the back. Ear mites (see pages 102–103) mostly affect the ears of young cats. They cause intense head shaking, ear scratching and a black, waxlike discharge, and can sometimes spread to the rest of the body, with generalised symptoms of irritation. Treatment is with topical parasiticidal drugs.

Skin Allergies

In many cases the cat's itchy skin is caused by an allergy. The cat then makes the problem worse by constantly licking and chewing the affected area so that the skin becomes inflamed and sore. The cat may be allergic to inhaled substances such as dust or house-dust mites, to certain foodstuffs, or to fleabites or chemicals that come in contact with the skin: some cats develop an allergy to the insecticide used in flea collars. Intradermal skin tests, blood tests, or hypoallergenic diet trials are usually necessary. Treatment will depend on the type of allergy, and may involve a diet change and anti-inflammatory agents.

Some Common Skin Problems

SIGNS	LIKELY CAUSE	ACTION
Black specks (flea dirt) in cat's coat or bedding. Scratching, chewing. Dull coat, skin appears scaly	Fleas/flea allergy	Treat with antiflea products to kill the fleas on the cat and to destroy the eggs and larvae in the environment
Excess scale (dandruff), some itching. Humans may have bites	Fur mites (Cheyletiella)	Treat with topical acaricidal (mite-killing) agents
Rubbing and scratching of ears. Black waxy discharge	Ear mites (Otodectes cynotis)	Acaricidal drops
Intense localised itching. Orange dots on feet, face and/or ears	Harvest mites (Trombicula autumnalis)	Acaricidal spray. Avoid mite-infested areas
Persistent licking and chewing. Inflamed skin	Allergy	Diagnose source of allergy. Symptoms may respond to corticosteroids and antihistamines

Poor Coat Condition

All cats, whether long or shorthair, should normally have a sleek, glossy and well-groomed coat. One benefit of grooming your cat's coat yourself is that it makes you familiar with its usual appearance and alerts you to any changes in its overall condition, such as areas of rough, scaly skin, bald patches, or lumps and bumps. If the coat suddenly looks dull and matted, it could suggest that the cat has an underlying health problem, particularly if this is seen in combination with other symptoms such as an alteration in appetite, noticeable weight loss, or vomiting.

Sometimes the problem may be as simple as a dietary deficiency, typically of a fatty acid, and can be corrected with supplements on your vet's advice. In an elderly cat a dull coat may be linked to overactivity of the thyroid gland, especially if the cat is eating more than usual. A blood test will quickly confirm the diagnosis. A cat that has problems with its teeth or mouth may stop grooming its coat altogether, which consequently becomes dull and unkempt. Check for signs of dribbling from the mouth, sore gums, or loose teeth, and take the cat to the veterinarian for a checkup.

▲ *Normal self-grooming by the cat helps to keep its coat in good condition and get rid of dead hair. Sometimes, however, a cat will groom itself so much that it licks the skin bare. This is often a sign of an allergic irritation or skin rash.*

Always check the cat's rear end when you are grooming – matting of the hair around the anus (possibly as the result of diarrhoea or a genital problem) can cause painful skin irritation. If the mats cannot easily be combed out, carefully cut away the lumps of hair with scissors, but be very careful not to cut the skin: seek expert assistance if unsure about doing this. Gently wash the sore skin with warm water, pat dry and apply a mild barrier cream. Have your veterinarian examine the cat as soon as possible.

Hair Loss

Cats shed dead hair throughout the year, which is why they need to groom themselves so constantly. Some (but not all) also have a heavy seasonal moult, usually in spring. Occasionally a cat may shed a large amount of hair as a result of a recent shock or emotional upset such as an acute illness, major surgery, or the arrival of a household rival in the shape of another cat or dog. The hair loss can occur several weeks after the event that triggered it, so it is not always easy to link the two. Some queens may lose hair during pregnancy and lactation. If your cat is constantly grooming itself but you are unable to find an obvious reason for it, you should consult your veterinarian. Sedatives or tranquillisers may be prescribed, or the vet may refer your cat to a behaviour therapist. In the majority of all such cases, the hair will normally regrow within a month or two.

Many cases of hair loss, however, are caused by scratching due to an allergy or the presence of parasites. Because cats are so secretive about their scratching habits, you may not have spotted the problem. Check the skin carefully for signs of fleas or flea dirt, a rash, or dandruff, and take appropriate action.

● Our elderly tortoiseshell cat, Flossie, has always had a magnificent semi-long coat. Over the last year it has gradually lost its shine. It seems dry and has started to mat. She seems to be grooming and eating normally. What can I do to improve things?

Are you feeding Flossie a balanced diet? A supplement of essential fatty acids, such as a combination of evening primrose oil and fish oil, can often improve a dull coat. These come in liquid or capsule form and can be purchased from your vet or a good pet shop. The dosage must be given daily for at least 2–3 months. You should have your veterinarian check Flossie thoroughly to make sure she hasn't a serious medical problem.

● After we got a new kitten from our local farm, our older cat, Bouncer, developed bald areas down his back legs and around his bottom. The kitten's coat seems fine. They haven't been that close to each other yet, so could it be something that Bouncer picked up from the kitten?

The kitten may have introduced fleas, which have infested the house and been passed on to Bouncer, who has a hypersensitivity to them. This is making him extremely itchy, and he is nibbling his hair out. Treat both cats for fleas with products recommended by your vet. Another possibility is that Bouncer may have reacted badly to the kitten's arrival and is pulling his hair out as a result of stress. Give him time and plenty of attention, and the hair should regrow. If neither of these seems to be the answer and the hair loss is continuing, consult your vet.

● We recently had to have one of our two cats put to sleep. Since then Omar, the surviving cat, has developed bald patches on his flanks. Could they be linked in any way?

Yes, stress caused by the loss of a close companion can make some cats behave abnormally. Often this manifests itself as excessive grooming, and the cat loses hair in symmetrical patches on either flank. Oriental cats are believed to be more susceptible than others. Consult your vet: if there is no underlying medical problem, such as parasites, you may be referred to an animal behaviour therapist.

▶ *This kitten has a ringworm infection on the top of its head. The exposed area of scaly skin is slightly raised at the edges. The fungus can be picked up from the soil and is passed from animals to humans.*

Ringworm

A common cause of patchy hair loss is ringworm (dermatophytosis), a highly contagious, fungal infection that seems to be most common in kittens and longhaired breeds. The patches formed by the fungus are often circular in shape, which give the condition its common name. The head, ears, paws and back are the most usual sites. Ringworm is easily spread by contact to other cats, dogs and humans, so if your cat has a suspicious-looking bald patch, visit your veterinarian without delay. Diagnosis may be confirmed by shining an ultraviolet lamp on the affected area in a darkened room as the fungus will sometimes fluoresce. The vet may also take skin and hair samples for microscopic examination and to grow a culture. The condition is treated with fungicidal lotions, ointments, or tablets, but it may take several months to clear completely. It is important to check other pets in the household to see if they are affected. Treating ringworm infection in a multicat household can be a major undertaking. You should disinfect all bedding, baskets and food bowls to make sure the infection doesn't spread. Don't let children play with an infected cat or kitten.

Skin Lumps and Swellings

OWNERS MOST FREQUENTLY NOTICE UNUSUAL swellings and bumps on the skin when grooming or stroking their cat. Skin lumps are quite common in cats and have many causes. They vary widely in size, feel and appearance. Some can be large and soft, others small and firm. Sometimes only one lump is felt; in other cases an area of skin may be covered in lots of tiny bumps. They may feel crusty, discharge pus, or ulcerate and bleed. Sometimes the skin is bald and scaling. Some lumps may be extremely itchy or causing obvious pain; others may appear to give the cat no discomfort at all.

Occasionally a tick on the skin is mistaken by the owner for a lump. These tiny parasites look like small, pale brown or grey pealike dots and swell rapidly with their victim's blood (see pages 88–89). They should be removed as soon as they are seen. Consult your vet if you are in any doubt about how to do this. Fur mats may also be confused with small skin bumps, especially in a longhaired cat that grooms and licks its coat excessively. Careful combing to remove tangles should prevent this (see pages 26–29).

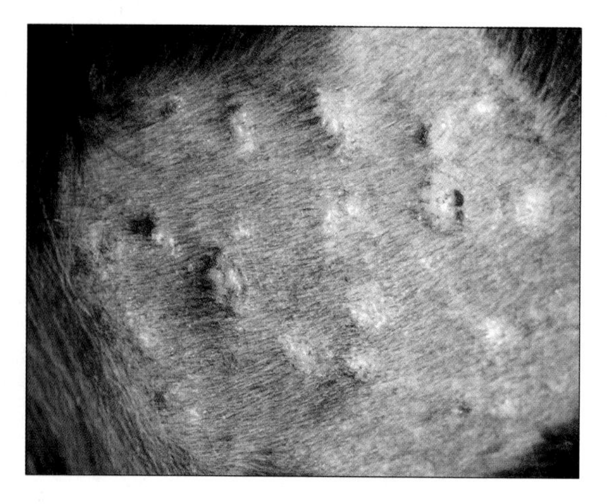

▲ *Multiple mast cell tumours are apparent on the skin of this cat (the area has been shaved). A biopsy (removal and microscopic examination of the affected tissue) is necessary to identify the nature of lumps like these.*

Identifying the Problem

The significance of an individual lump can be difficult for an owner to assess. No lump should be dismissed lightly, so it's best to have your vet examine it as soon as possible. The vet is likely to require a detailed history from you (see box). Some lumps can be diagnosed on the basis of examination and history alone. For example, a cat-bite abscess resulting from an infected puncture wound (see Fight Injuries, pages 130–131) will appear a few days after the cat has been in a fight. You will notice a swelling on the cat's skin, usually in the region of the face or tail, which feels warm to the touch and is causing obvious pain. The cat may seem subdued and off its food. If the abscess is on a limb, it may be lame. Sometimes the abscess bursts, releasing pus. If it does not discharge spontaneously, the vet will drain it surgically and prescribe antibiotics.

Other lumps and tumours can be more difficult to identify. Small, crusty bumps and inflammatory swellings may be the result of an allergic reaction to parasites, especially fleas, or a food hypersensitivity. To pinpoint the cause, the vet will analyse skin and hair samples and carry out allergenic tests. Sometimes there is no identifiable trigger, but the symptoms may be treatable with anti-inflammatory medicines. Other conditions that may lead to skin lumps and swellings include cysts and tumours, fungal, bacterial and viral infections, or an embedded foreign body such as a thorn or gun pellet. In many cases a biopsy and other laboratory tests are needed to arrive at the root of the problem.

Cancerous Lumps

In some cases a biopsy will establish that a lump or tumour is cancerous. The most common kind of skin cancer are squamous cell carcinomas. They affect areas of skin most exposed to the sun's radiation (the ears, eyelids, nose and lips) and are most common in cats with white coats and large areas of depigmented skin. Initially the

What To Tell Your Vet

To aid diagnosis, the vet is likely to ask you for a full history and description of the lump or lumps. Be prepared with your answers.

- When and where did you first notice the lump?
- Has it grown since you first spotted it? If so, how quickly?
- Is the cat off its food or otherwise unwell?
- Has it been in a fight recently?
- Is the cat scratching and licking the area?
- Is it up to date with its antiflea treatments?
- Is the lump painful? Does the cat respond aggressively when you touch it?

• The central pads on our cat Pushkin's front paws have become soft and swollen. He seems fine otherwise. What's the cause?

From your description, it sounds like a disease known as plasmacytic pododermatitis. This is uncommon and is thought to have an immune basis: it can even be seasonal in some cats.The pads may spontaneously return to normal, or they may ulcerate and cause lameness. You should take Pushkin to your vet – diagnosis will require a biopsy. Treatment is not always necessary, but if the symptoms are serious, surgery or anti-inflammatory therapy can help.

• I've noticed that Osmin, our tom, often has a swollen chin. He occasionally gets fleas. Could there be a connection?

I suspect Osmin has what is loosely termed 'fat' or 'lumpy' chin, an inflammatory disease that tends to be cyclical and is thought to be caused by an underlying allergy, either to fleas (as seems possible in this case), food, or inhaled allergens. Sometimes no cause is found. Treatment is with anti-inflammatory drugs and control of the underlying cause.

▶ All white, white-faced, or white-eared cats are vulnerable to the sun's rays, which can cause skin cancer. To be on the safe side, keep your cat indoors on very sunny days.

skin appears reddened and sunburned – the precancerous stage. Then scaling and small lumps develop, which crust and ulcerate. It is a malignant cancer and may spread. Treatment involves removal of the affected tissue, possibly the whole earflap. This radical surgery is often curative, but sometimes further treatments are needed.

Sometimes an owner discovers a lump in the mammary glands (breasts) of an older female cat. Breast lumps appear in both spayed and unspayed cats and can vary from a small solitary nodule to a diffuse plaque, or patch. Any lump along the mammary area should be investigated without delay. In unspayed females a very high proportion of such tumours are malignant (the risks are lower in cats spayed before their first heat). Mammary cancer spreads rapidly and may reoccur after treatment, but early treatment (surgical removal and sometimes chemotherapy) gives a much better long-term outlook. Some mammary lumps turn out to be harmless cysts. They are the commonest type of cyst in cats and may result from prolonged hormone treatment.

Eye Problems

THOUGH BASICALLY SIMILAR IN STRUCTURE TO the human eye, the cat's eye nevertheless has a number of significant differences. Cats need only a sixth as much light as humans do to be able to make out shapes and movement. Extra curvature at the front of the large eye picks up the maximum amount of light, and a special covering at the back (known as the tapetum lucidum) acts as a mirror and reflects any unabsorbed light back onto the retina, the light-sensitive membrane that receives the image. The cat's eye also has a more elastic lens than ours. This allows them to alter their range of vision more rapidly than we can, though their close-range vision is less good.

Another feature of the cat's eye is the nictitating membrane, a 'third eyelid' in the inner corner of each eye that acts as a protective screen in case of damage to the surface. If the third eyelid is visible, it may indicate that the cat has an eye injury, is suffering from dehydration, infection or possibly stress, or is sedated. Any changes you notice in your cat's eyes should be investigated by your vet as soon as possible. Look out for signs such as reddening, a half-closed eye, changes to the pupil, or a heavy, discoloured discharge (see box on page 100).

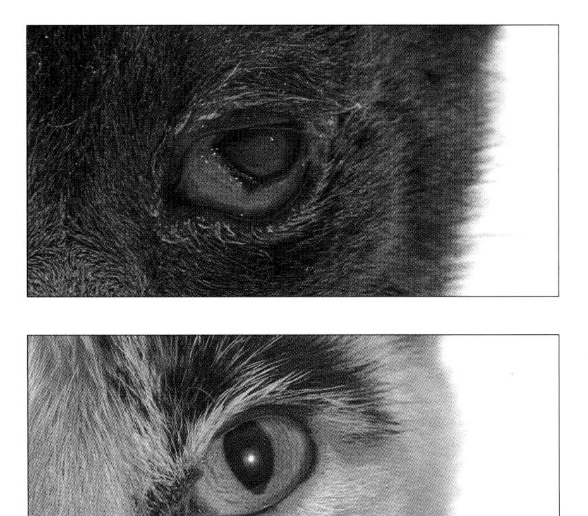

▲ *Conjunctivitis (top), an inflammation of the outer layer of the eye, may be due to local infection, allergic reaction, or a viral disease. (Bottom) A blocked tear duct leads to runny eyes with buildup in the corner.*

ANATOMY OF THE EYE

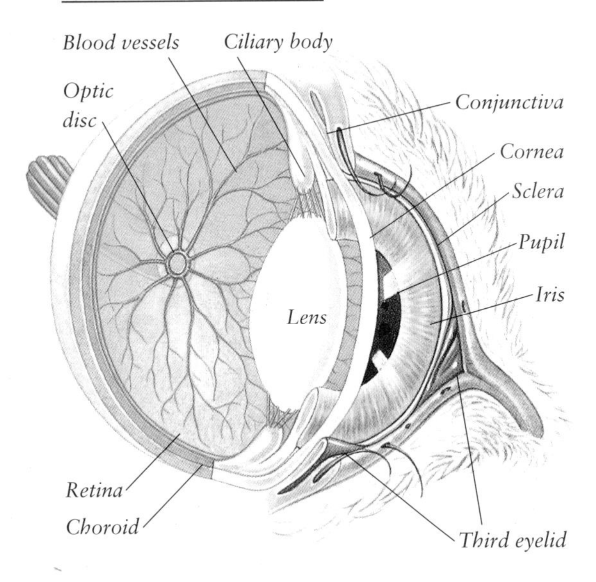

▲ *The structure of a cat's eye is similar to ours, though much larger in proportion to its head. Fluid (the vitreous humour) gives it shape. Most injuries occur on the surface, especially to the eyelids and cornea. If the 'third eyelid' is visible, it is a sign of stress or illness.*

External Problems of the Eye

Fighting cats will literally try to 'scratch each other's eyes out', and this is the most common kind of injury. A half-closed eye with fluid pouring out of it is an emergency and must be treated by the vet immediately; if the injury is severe enough, the whole eye can be lost. More commonly, there may be scratches around the eye or on the cornea (the surface of the eye), causing watering, blinking, or redness. Always have fight injuries to the eyes treated to prevent a corneal ulcer from developing or infection at the site of the wound spreading to deeper parts of the eye.

Conjunctivitis is a condition in which the conjunctiva (the pink fleshy membrane surrounding the eyeball) becomes inflamed, swollen and reddened as the result of allergy, irritant, flu virus, or another infectious agent. It occurs in one or both eyes and is often accompanied by a clear or yellow-green discharge. Your vet will probably prescribe antibiotics and eye drops to clear the condition up. You should never give a cat eye drops intended for humans.

Longhaired breeds, especially those that have the squashed faces and prominent eyes typical of Persians and their crosses, are prone to develop blocked tear ducts. This condition is not painful in itself (unless it is caused by an infection) but the cat's eyes should be bathed regularly to prevent the surrounding fur from becoming matted. Blocked tear ducts can be flushed out by a vet. It is usually necessary to anaesthetise the cat in order to carry out this procedure.

● ***Is there any way I can tell if my cat is going blind?***

A... You can test the sight of a kitten or young cat by dangling a paper or cotton ball on a string in its line of vision and watching to see if it follows the motion with its eyes or plays with the ball. Signs of failing sight or loss of vision in an older cat are moving more slowly than usual, staying close to the wall when walking through the house, misjudging the height of jumps, or bumping into furniture. If you suspect the sudden onset of blindness, see your vet without delay – it may be possible to restore the cat's sight if treatment is given promptly.

● ***I have a lovely white Angora, Snow, who is very clean but usually has a brown patch around her eyes. Is this something to be concerned about?***

Most cats have a clear discharge from the corner of their eyes, and occasionally this leads to a brown discoloration on the face of white cats. If Snow has no other symptoms and the amount of discharge does not increase, there is probably not a problem. You can clean the area gently with damp cotton wool to help her keep it clean. If it gets worse, consult your vet.

What To Tell Your Vet

- When you first noticed the problem.
- Whether a new cat has recently arrive in your household.
- How many of your cats have the problem.
- If there was any obvious pain or irritation at the beginning.
- Whether one or both eyes have been affected, and whether at the same time or separately.
- The colour and the consistency of any discharge from the eye.
- Whether your cat could have been involved in a fight recently.
- Whether your cat has had all its vaccinations or annual boosters.
- Any other signs of ill health: loss of appetite, general listlessness, etc.

▶ *This kitten has suffered an injury to the right eye. The vet has applied a special stain to the eye, which will indicate if the cornea has been scratched or ulcerated as a result.*

Common Eye Symptoms

Signs	Possible cause	Action
Surface looks cloudy ('blue')	Cornea inflamed due to injury (often a fight) or infection	Wipe eye gently with damp cotton wool. Keep out of bright light, and see your vet
Pupil looks cloudy or white	Lens cataract – typically an age-related change. Can also be due to injury, infection, diabetes, or congenital defect	See your vet to check cat's general health
Fixed, dilated pupils	Retinal problems, glaucoma, cataracts, tumours, infections of the eye or brain, or a blow to the head	See vet as soon as possible
Eyeball protruding from socket	Prolapsed eyeball – usually trauma-related, but may also be due to tumour at back of socket	Cover with moist, clean eye pad. See vet immediately
Eye looks red	Haemorrhage (blood) in the eye, conjunctivitis	See vet as soon as possible
Prominent third eyelid ('skin across eye')	Trauma to eye, general poor health, dehydration after diarrhoea or vomiting, sedation	Monitor cat's health. See vet in 24–48 hours if no better
Eye half-closed	Foreign body in eye, irritant, flu, trauma (injury)	Clean eye gently with damp cotton wool every hour. If no improvement, see vet
Clear discharge	Blocked tear ducts, allergic conjunctivitis	As above
Sticky, yellow, or green discharge	Infectious conjunctivitis, tear film problem	As above. Check vaccine status. Keep isolated, see vet

Eye Diseases

Cats can suffer from three serious diseases of the eye, all of which can progress to blindness. Progressive retinal atrophy (PRA) is a degenerative disease of the retina that occurs in both pedigree and nonpedigree cats. The condition is hereditary in Abyssinians and most typically affects young kittens within the first three months of life. There is a higher than average incidence in Siamese, but it affects this breed in middle age, and no genetic link has been proven. PRA is not painful but is progressive and untreatable, leading eventually to total blindness.

Cataracts are most often seen in elderly and diabetic cats but they may also occur in younger cats as a result of a congenital defect. An opaque 'cloudy' patch appears on the lens and gradually spreads across it if left untreated. Sometimes an elderly cat may have an opacity of the lens that is not a cataract, but it is always worth having the cat checked by the vet. Surgery may be effective in restoring some vision to the eye.

Glaucoma is relatively uncommon in cats, but it occurs if too much fluid pressure builds up within the eye. The excess fluid causes the eyeball to enlarge and exert pressure on the delicate retina, damaging the cells and causing pain as well as reducing vision. Glaucoma can be treated with eyedrops, pills, or surgery if it is caught early enough; if not, permanent blindness is the result. Enlarged eyeballs, fixed pupils, a staring expression and a cloudy cornea are the usual symptoms of this disease.

Following surgery or injury, the third eyelid is often closed and sutured in place to allow the cornea to heal underneath. Alternatively, clear contact lenses may be used as a form of bandage.

▶ *To give your cat eye drops, use one hand to hold your cat's head and apply the drops with the other. Do not point the dropper or tube at the eye – hold it above or parallel to the eye and squeeze. When a drop has fallen onto the eye, hold the cat's eyelids together gently for a minute to spread the ointment inside the closed lids.*

Our longhaired calico, Pumpkin, goes around squinting and is always rubbing one eye. It looks as if her eyelid isn't normal. Can we do anything about this?

Pumpkin may have entropion – a condition in which the edge of the eyelid (one or both) is rolled inwards so that the lashes rub against the surface of the eye, causing squinting, irritation and discharge. The eye must be examined by a vet. If entropion is responsible, it can be corrected by surgery.

My old shorthaired male, Ram, is 17. He is totally blind but otherwise healthy. I can't bear the thought of putting him to sleep, but is it cruel to let him live longer in this condition?

No. With their senses of hearing and smell intact, cats can adapt to blindness without too much distress as long as their environment is stable and secure. Unless he is in pain or being attacked by other cats for staring, Ram should be content.

▲ *A button holds down stitches to keep this kitten's third eyelid closed, allowing the cornea to heal underneath. It causes no discomfort and prevents the kitten from removing the stitches by scratching at them.*

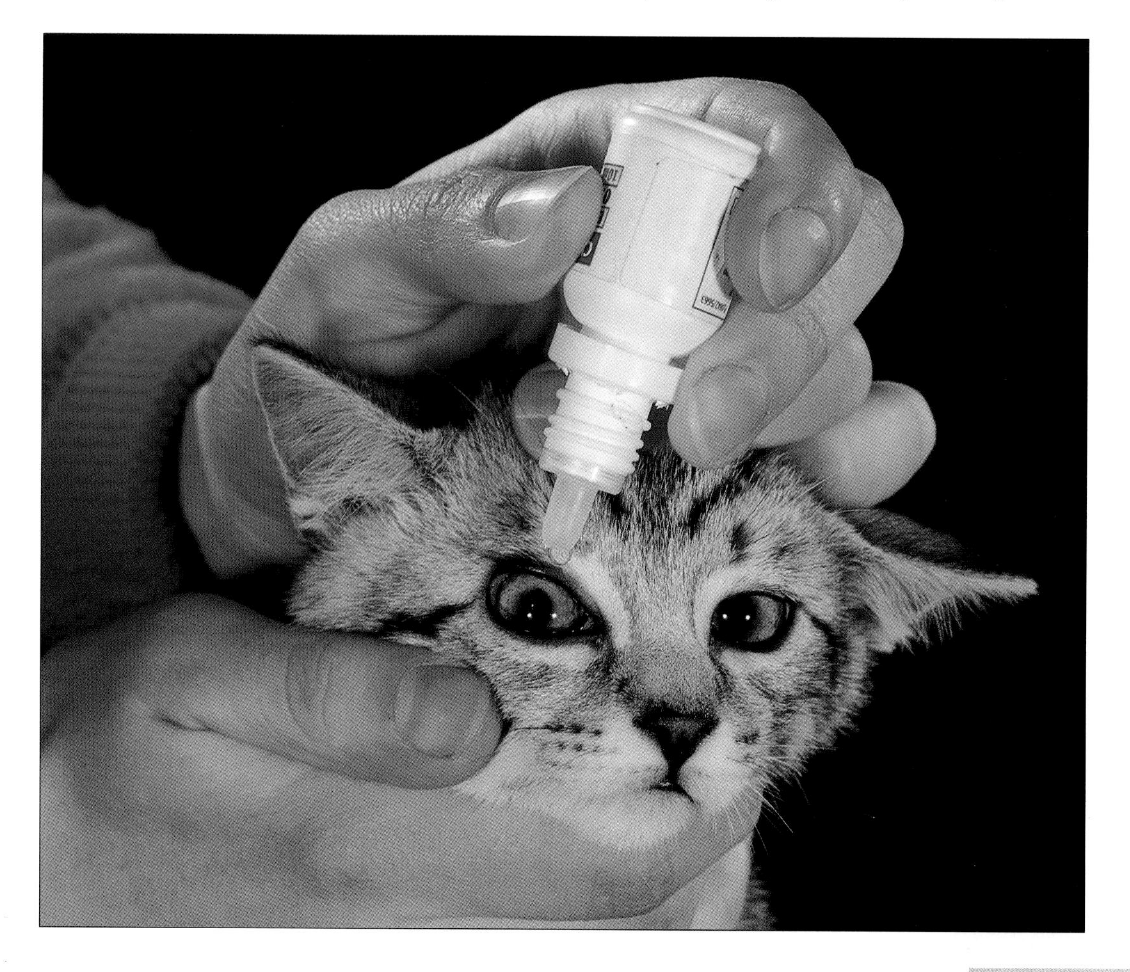

Ear Problems

A CAT RELIES ON HEARING AS MUCH AS SIGHT for hunting and survival in the wild. Its erect and mobile ear flaps are designed to locate and pick up noises easily. The sound waves travel down the ear canal to the ear drum, a thin membrane that vibrates when stimulated by sound waves. The vibrations then pass through the three tiny bones of the middle ear to the cochlea, within the inner ear, which converts the sound vibrations into messages that pass along the auditory nerve to the brain. Also in the inner ear are the semicircular canals, the fluid-filled organs that control balance (see pages 126–127).

A large echo chamber on either side of the cat's skull make it highly sensitive to particular frequencies of sound, especially the high-pitched squeaks and rustling noises made by small prey animals. A cat is able to hear sounds as much as two octaves higher than humans. Like humans, however, its sensitivity to high noise may decline as it begins to age. This may become noticeable by the time the cat is around 4 years old.

What Can Go Wrong

Because they are so prominent, the cat's ear flaps are easily torn, bitten, or scratched in cat fights. Deep wounds should be examined and treated by a vet to avoid the risk of infection. Cats with white or white-tipped ears are particularly prone to sunburn, which can lead to skin cancer (see pages 96–97). If you see any sign of reddening, swelling, or hair loss, lose no time in having the vet investigate it; these conditions could be precancerous. If a cancer has already developed, it may be necessary to remove some or all of the ear flap to prevent the cancer spreading to the head. This will not affect normal hearing.

Frequent head shaking and scratching of the ear, especially in kittens, may be a sign of infestation by ear mites. These tiny parasites are normally present in small numbers but sometimes multiply, causing extreme irritation. There may be a discharge of brown wax. Treatment is with parasiticidal ear drops; all cats and dogs in contact with the infested cat should be dosed (see pages 88–89). Also treat the neck area with a flea powder for cats, to prevent mite eggs from sticking to the fur and causing reinfestation.

Fungal and bacterial infection of the ear, a foreign body such as a grass seed, and inflammatory polyps or tumours of the inner or middle ear may cause symptoms (see chart). Have any ear problem investigated without delay, as an untreated infection may lead to deafness. The vet will examine the ear with an otoscope, an illuminated tube with a magnifying lens, and may also take X-rays. The cat may need a sedative or anaesthestic. Infection is treated with antibiotics and anti-inflammatory drugs.

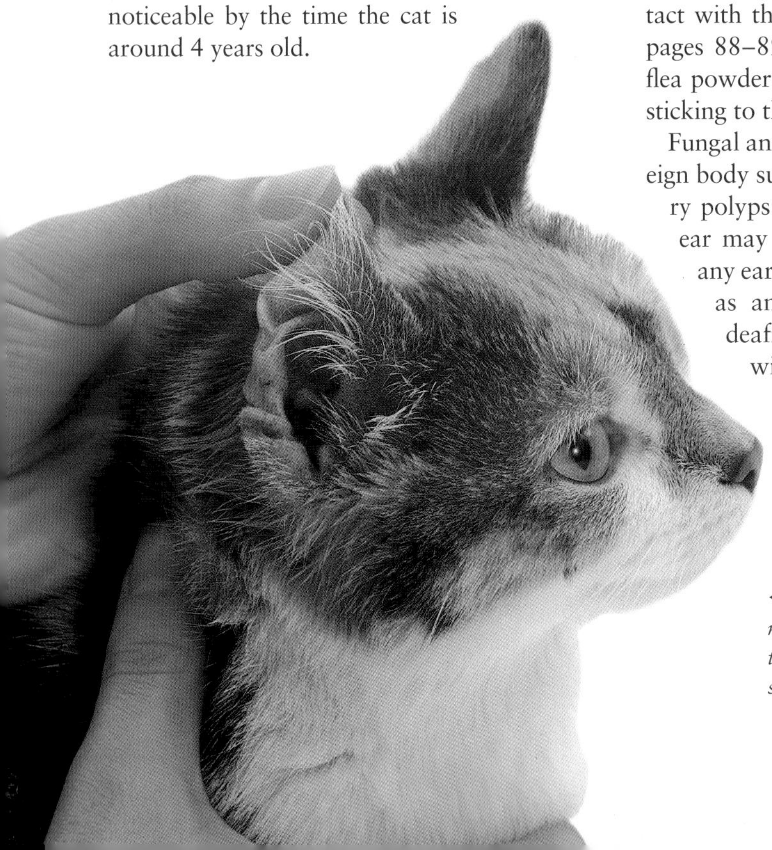

◀ *This young cat is infected with ear mites. The inside of the ear looks red, and there is a brownish discharge. It has been scratching, making the ear sore.*

Some Common Ear Disorders

Signs	Possible cause	Action
Holding down the ear, vigorous head shaking	Fight injury	Clean the wound with bactericidal soap or antiseptic cream. If wound is deep, have the vet examine it
Reddening and swelling of ear tips	Sunburn	Examination and treatment by the vet
Raw and bleeding ear tip, constant scratching of ear	Squamous cell carcinoma	Removal of all or part of ear flap
Head shaking, ear scratching, waxy discharge	Ear mite infestation	Treat with ear drops. Treat fur around neck area with flea powder
Head shaking and ear scratching with audible fluid sound	Tumour or polyp in middle or inner ear	Surgical excision
Head shaking and tilting, rubbing and scratching ear, obvious pain, discharge	Fungal or bacterial infection	Treat with antibiotics and anti-inflammatories
Loss of balance, head tilting, walking in circles, eye flickering	Inner ear disease (see pages 126–128)	Take cat to the vet without delay

● ***Our new crossbred kitten, Millie, has very dirty, itchy ears. The vet cleaned them out two weeks ago and gave us some drops to use. She now seems fine, but the vet has told us to continue putting the drops in for another month. Why is this?***

Millie's ears are infested with ear mites, and the vet has prescribed ear drops to kill them. However, the medicine kills only the mites themselves, not the eggs they have already laid inside the ear. These may take up to 6 weeks to hatch. Consequently, the drops should be given for this length of time. Also treat Millie's neck for fleas to prevent reinfestation from any mite eggs that may be sticking to the fur there.

● ***Jade is 11 months old and has been shaking her head and scratching her ears for several weeks. Recently her left ear has become swollen. The vet says the swelling is blood-filled and requires drainage. What is it exactly?***

This is an aural haematoma, or blood blister, a fluid-filled swelling of the outer ear. The fluid often looks like watery blood when it is drained. It is thought to result from ruptured blood vessels in the ear caused by the cat constantly scratching the area and shaking and rubbing its head, possibly because of an infected wound or parasites. Drainage is the usual treatment. If left untreated, the condition may lead to scarring and a 'cauliflower' ear.

▲ *A white cat with one blue eye and one of a different colour is quite likely to be deaf in one ear.*

Deafness

Deafness is inherited in some cats. Mainly white cats are affected, especially those with blue eyes. Sometimes, if one eye only is blue (odd-eyed), the cat may suffer from one-sided deafness, but this is not usually a problem. Cats with inherited deafness can be difficult to train. It is advisable to keep them indoors as they will be unable to hear traffic noises; similarly, an elderly cat that takes to sitting in the road may have become too deaf for its own safety. But if it is provided with plenty of company, there is no reason why a deaf cat cannot lead a full and happy life.

Dental and Mouth Problems

In the care and grooming of our cats, one of the areas most likely to be neglected is the mouth. If you do not look inside it regularly, you may miss important signs that your cat is suffering discomfort or even severe pain from periodontal disease or some other problem. If your cat seems to resent any effort to touch or open its mouth, something may be wrong. Are the teeth intact and white? Are the gums pale pink and evenly coloured? Does your cat also protest at having its teeth or face touched? Are there any sores on the palate or the tongue? Does the cat's breath smell foul? Does your cat drool and paw at its mouth? If you notice any of these things, make an appointment with your vet promptly.

▶ *A cloth is used to clean a kitten's teeth. Oral hygiene is best begun early, teaching the cat to tolerate the activity.*

The Roots of Periodontal Disease

Your cat's teeth should be cleaned for the same reason that you brush your own. Bacteria forms plaque on the teeth and lodges between them and the gums. As it accumulates, inflammation of the gums occurs. This is called acute gingivitis. If it becomes serious enough, your cat may develop an infection requiring antibiotics.

Underlying diseases, such as a kidney condition, diabetes, or viruses, can cause chronic gingivitis. This condition is even more painful, with extensive ulceration of the gums and other areas in the mouth. Treatment is difficult, and it may even prove necessary to extract all the teeth. Bacterial infections may eventually invade the space between the root of the tooth and its socket in the skull or lower jaw. At this point, periodontal disease sets in and progresses until it destroys the bone of the socket and the supporting fibres that hold the tooth in place. The tooth becomes gradually looser, and sooner or later it falls out completely. Periodontal disease is often accompanied by oral abscesses and pain, especially when eating.

Over 60 per cent of cats suffer from a condition called feline resorptive lesions or neck lesions. These are cavities that appear at the 'neck' of the tooth, the area at the gum between the root and the crown, but are dissimilar to the bacterial decay (caries) found in humans, which is rare in felines. The cavities eat away the tooth and eventually right through into the pulp canal. Ultimately the whole crown disintegrates. At the

Preventing Dental Problems

Toothache and a sore mouth can make it difficult for your cat to eat or groom itself. To prevent problems, you should:

- ✓ Feed your cat good-quality dried food or give it something hard to chew.
- ✓ Brush your cat's teeth once a week or have them descaled by the vet, as needed .
- ✓ Check your cat's mouth regularly for sore red gums, brown tartar, loose teeth, bad breath.
- ✓ Have your cat's mouth examined by your vet at least every 6 months, perhaps when it goes for a regular checkup.
- ✓ Ask your vet to investigate any painful areas on the tooth, lumps in the mouth, or ulcers on the lips, gums, or tongue.
- ✓ Take your cat for an exam if it starts pawing at its mouth, drools, or has obvious difficulty in eating.

● ***How can I get my cat to tolerate having her teeth brushed?***

... Starting her as a kitten will help, but if she is already an adult, it may not be too late. Use cotton buds to touch her teeth lightly, getting her used to the activity over several days. Put a little special cat toothpaste on your finger and let her lick it; then put a tiny dab on her mouth. After a few weeks try the first brushing, using your finger or a finger brush (available from pet shops). Eventually you should be able to use a small soft toothbrush.

● ***My cat keeps salivating and pawing at his mouth. What can be the matter?***

Open his mouth and have a look. Check for foreign bodies in the mouth: a needle sticking in his tongue or a fish bone between the teeth, for example. Check the condition of his teeth and gums, and look for loose teeth or ulcers on the roof of the mouth or tongue. If you can't solve an obvious problem, you should take your cat to the vet as soon as possible.

● ***My vet says that my cat's teeth are so bad that they must all come out. How will she eat?***

The answer is 'much better'. Your cat will develop very hard, callused gums and will even be able to manage dry food. Crunching against hard gums is considerably more comfortable than biting down on painful teeth.

same time, similar lesions may be eating into the root. The cause is unknown, and no treatment is very effective except extracting affected teeth. Meanwhile, the cat suffers severe toothache.

Cats also suffer from a number of diseases that affect soft tissue in the mouth. Sore gums and ulcers may be a sign of a viral disease such as leukaemia or an auto-immune disease. Particularly alarming to look at are eosinophilic granulomas (rodent ulcers), large lip ulcers, or raised ulcers on the tongue, the roof of the mouth, or the throat. They are probably caused by an allergy and can be treated by your vet.

Most worrying of all are tumours that occur in cats' mouths. These can be malignant and must be removed early for treatment to succeed. This makes regular dental examinations essential.

How to Clean Your Cat's Teeth

Clean your cat's teeth at least once a week (twice if your cat has already developed dental problems). Wrap the cat in a towel to restrain it. Get someone else to hold it firmly and pull back the upper lip. Gently pull down the lower jaw and rub the teeth with a finger brush, an extra-soft toothbrush, or a piece of cloth dipped in a weak salt solution or smeared with a special animal toothpaste – never use toothpaste meant for humans. If your cat struggles too much for you to carry out cleaning at home, take it to the vet once a year to have its teeth descaled.

▼ *This cat's mouth badly needs treatment. Dental examination and treatment of cats must almost always be carried out under general anaesthesia.*

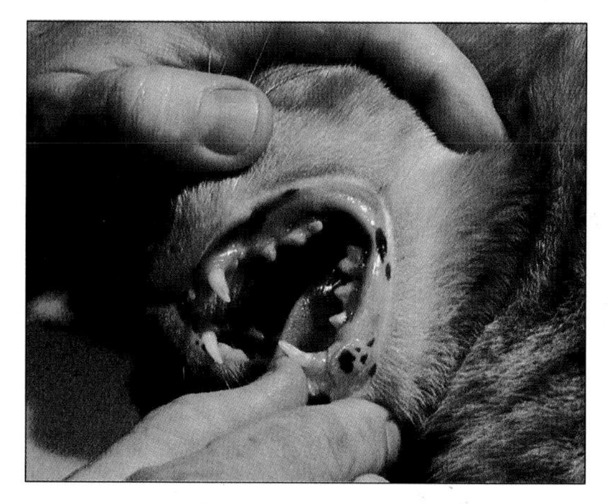

Digestive Problems

VOMITING FORMS PART OF A CAT'S NATURAL defence system, enabling it to eliminate excess or harmful foodstuffs. It is quite normal for some cats (particularly those that hunt and eat their prey) to vomit once or twice a week without any obvious ill effects. Grass-eating, overeating and fur balls may also cause vomiting but should not normally cause concern. Sometimes a kitten will bring up roundworms, a sign that its deworming treatment has been neglected. However, if your cat begins to vomit more frequently than usual, and if the vomit contains blood (resembling coffee granules) or the cat is losing weight, there could be an underlying problem such as a bowel obstruction or tumour. You should have the vet investigate the problem without delay.

Acute Vomiting

Various forms of gastritis can cause acute attacks of vomiting. If your cat throws up two or three times within a short period of time, confine it to the house and give it small amounts of water. Should the cat vomit up the water, restrict its fluid intake for 8–12 hours to allow the stomach to rest, then give it a tablespoonful of water every 30 minutes – it is important not to allow the cat to become dehydrated. You should withhold all food until there has been an interval of at least 8 hours since the last vomiting attack, then offer a small amount only and carefully monitor progress. If the cat is still vomiting and cannot keep down water, take it to the veterinarian without delay.

Unvaccinated kittens and cats are at risk from feline infectious enteritis (see pages 86–87). In the more acute form of the disease, the cat suffers serious and frequent attacks of vomiting, diarrhoea and abdominal pain. Feline infectious enteritis is often fatal. It is also highly infectious, and the best form of prevention is vaccination.

Very occasionally vomiting may occur after a cat has swallowed a small object such as a piece of cloth or plastic that causes a blockage in the gut. If this is suspected, X-rays, endoscopy, or surgical exploration may be required to locate and remove the object. The problem is much less common in cats than in dogs.

Sometimes a cat will regurgitate food before it has reached the stomach, often in a compacted 'sausage' shape. The cat may re-eat the regurgitated food, something that many owners find disgusting. Occasional regurgitation is often due to the cat's eating too quickly and is no cause for concern. However, if it occurs persistently, ask your vet for advice; there may be a problem.

▼ *A permanently visible third eyelid is a sign that your cat is unwell. It can be caused by dehydration following an attack of diarrhoea, among other disorders.*

What To Tell Your Vet

Vomiting

- How long after feeding or drinking did the vomiting occur?
- How frequent is it?
- Is the cat keeping down water, when offered?
- What colour and consistency was the vomit? Did it contain identifiable food, evidence of worms (white threads), or fur?
- Was there any sign of blood?
- Have you noticed any other symptoms such as diarrhoea, weight loss, change in appetite?
- Has the cat's diet recently been altered, or has it eaten anything unusual?
- Is it receiving any medication at present?

Diarrhoea

- When did you first notice the diarrhoea?
- How often is it passing a stool?
- What colour and consistency is it?
- Is any mucus or blood present?
- Does the cat strain to pass a stool or show signs of pain?
- Have you noticed any other symptoms such as vomiting or oversleeping?
- Is the cat eating more or less than usual?
- Do you give the cat milk to drink? Has its diet altered recently?
- Is it receiving any medication at present, including deworming or antiflea treatments?

● ***Our cat Ziggy seems to like to chew grass and is often sick afterwards. Is this unusual?***

No, cats will quite commonly do this, though the reason for the behaviour is still unclear. It certainly does them no harm, provided the grass has not been sprayed with chemicals. Cats that live permanently indoors may even appreciate a small tray of cultivated grass to nibble.

● ***Dorabella, my Siamese, is a bad car traveller. She dribbles profusely and often throws up. How can I help her?***

Don't feed her before a journey and cover her carrier with a towel. Frequent short trips, or putting her in the car for short periods when stationary, may help acclimatise her to it. If this fails, discuss with your vet the use of a sedative for longer journeys.

● ***Monty, our 2-year-old ginger cat, loves to eat liver once a week. The last couple of times I've given it to him I have noticed staining on the fur around his bottom afterwards, as though he has had diarrhoea. Should I continue to give him his treat?***

Liver is a rich food for cats, and it seems likely that it is causing Monty a mild dietary upset each time he has it. Leave it out of the diet for now. If the diarrhoea ceases, try reintroducing an occasional small liver snack, but if it recurs, stop giving him liver altogether. If the liver clearly wasn't the culprit and the diarrhoea continued after he stopped having it, you should have your vet examine him.

Looseness of the Stool

The passing of unformed or liquid faeces can indicate a range of digestive and dietary problems. They include overfeeding, a change of diet, the introduction of rich foods and worm infestation. Some kittens have an intolerance to lactose in cow's milk, and this can lead to loose faeces. Any kitten with diarrhoea lasting more than 48 hours should be examined by a vet, even if the kitten seems to be generally healthy.

Cats are very private about their toilet habits, so if your cat does not use a litter box, you may have difficulty in knowing what is a normal stool for your cat and how frequently it defecates. One sign that your cat has a diarrhoea problem is staining of the fur around its bottom.

If your cat has a sudden attack of diarrhoea, keep it indoors and give no solids for 24 hours. Provide fresh water only. At the end of that time, begin to offer it a bland diet – baby food, home-cooked white meat or fish, or a commercially prepared food of that type – given in small amounts 4 to 6 times daily. Give your cat about half a normal day's food in the first 24 hours, increasing the amount gradually if no further diarrhoea is seen. Continue like this until the cat has passed normal stools for 48 hours. If the diarrhoea lasts any longer, or the cat is passing blood with the faeces, see a vet without delay.

Weight loss accompanied by chronic diarrhoea may be a sign of a hormonal disorder, such as an overactive thryroid gland. In this case the cat is likely to develop a ravenous appetite. The condition most commonly affects older cats. Your vet will carry out a blood test to check the cat's thyroid hormone levels. In most cases surgical removal of the thyroid gland is successfully carried out. Black, tarlike faeces may indicate internal haemorrhaging or a tumour. It should always be taken seriously.

If possible, take a recent sample of the cat's faeces with you for testing to aid the vet's diagnosis. To obtain a sample, you may have to keep the cat indoors with a litter tray for a day or two. Collect the sample in a clean container and label it with your cat's name, the date and the time it was collected.

Food Sensitivities

Many cats have a sensitivity to a particular food such as liver, or to more than one food. Milk should never be given instead of water to drink, but regarded as a food. If your cat has chronic diarrhoea, remove milk and dairy products from its diet altogether. Minimise diarrhoea in a cat that is known to have food sensitivies by excluding all home-cooked foods and giving it a good-quality commercial diet. Your vet will advise you. Watch carefully to make sure your cat does not scavenge or beg food from other sources, such as kindly neighbours, as this will interfere with any dietary control you impose.

Constipation

Constipated cats will pass only a hard, dry stool or none at all. After repeated straining, the cat sometimes passes liquid faeces. If the cat holds its faeces in too long, more fluid than usual is reabsorbed by the gut, and the resulting mass may be too solid to pass. Cats dislike using soiled litter trays, and this may cause them to hold their faeces in. Stress, fur balls, dehydration, tumours, or an obstruction to the bowel (caused perhaps by scavenging for chicken bones) can all lead to constipation. It is also common in older cats.

You will become aware of the problem when you fail to observe the cat defecating for several days or notice it straining unproductively. This can be confused with straining

◀ If your cat is suffering from constipation, the vet will carry out a full physical examination, including palpation of the abdomen (as here), a rectal examination, and possibly an X-ray or ultrasound.

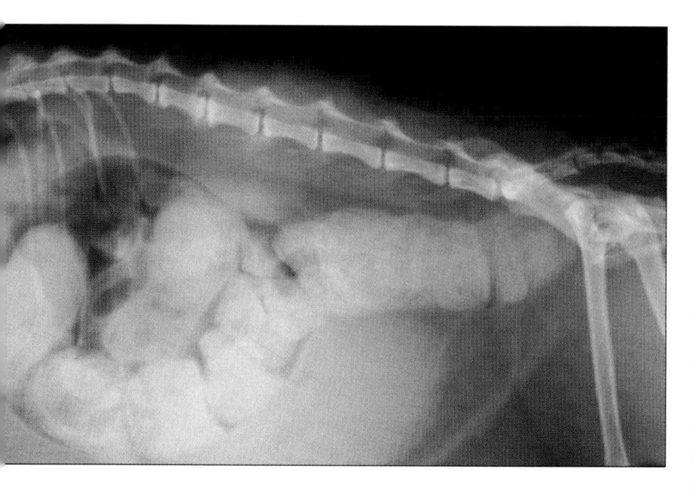

▲ *This X-ray clearly shows a buildup of faeces in the bowel of a constipated cat. You can help avoid constipation by adding fibre to your cat's food and grooming it more frequently to prevent fur balls.*

to pass urine (in cases of cystitis), so observe carefully. The cat may use the litter box just to urinate or pass only small, dry faeces. If the constipation continues for several days, the cat will gradually become weak and dehydrated, and it may begin to vomit. Treat this as a medical emergency, and get the cat to a veterinarian without delay. The vet may try to remove the faecal mass by administering an enema and/or a laxative. Sometimes this is done under general anaesthetic. Faecal softeners or expanders (such as bran) may be needed to bulk out the faeces. The vet may prescribe a therapeutic diet for a few days afterwards or advise you to give small amounts of digestible diet such as fish, chicken, bran, or lightly cooked egg.

Abdominal X-rays, an ultrasound scan and blood tests may be carried out to investigate the cause of the problem, monitor the cat's progress, and determine appropriate treatment. Depending on the underlying condition, your cat may need a special diet for the rest of its life. A condition known as megacolon, common in younger cats (2–9 years), leads to pocketing of the faeces and dilation of the colon. Drugs can be given to promote the passage of faeces through the gut, but it may sometimes be necessary to remove the affected portion of the colon.

● ***Sophie, my Ragdoll, recently suffered a pelvic fracture in a road accident. The vet has advised me that she is likely to be constipated as a long-term consequence and should be given laxative medicine regularly to prevent it. Could this have any long-term effects on Sophie's health?***

Yes, the long-term use of laxatives can interfere with the absorption of fat-soluble vitamins (vitamins A, D, E and K), which are essential nutrients. You should therefore make sure that you give Sophie a regular vitamin supplement – your veterinarian should be able to advise you.

● ***The vet has recommended giving Fingal, my ginger tom, a daily dose of mineral oil for his constipation problem, but I have tremendous trouble administering it – Fingal can be very independent and uncooperative. Do you have any advice?***

Always take care when giving a liquid medicine by mouth – if the cat inhales any into its airways it can lead to a severe type of pneumonia. Give the mineral oil in small amounts only (from a teaspoon or pipette), allowing plenty of time for him to swallow one dose before giving the next. Flavouring the tasteless mineral oil with something tasty like sardine oil can help, or ask your vet to recommend one of a number of attractive-tasting laxatives that are available. Alternatively, you can try wiping petroleum jelly on the cat's whiskers or paws, to be licked off as it grooms.

What To Tell Your Vet

CONSTIPATION

- Does the cat repeatedly strain to pass faeces and/or appear to have pain?
- Is its abdomen enlarged and tense?
- Does the cat groom itself excessively? Fur balls often cause constipation.
- Has it recently been hospitalised, stayed in a cattery, or undergone any other kind of stressful change to its lifestyle?
- Does the cat scavenge for food or hunt prey (could it have swallowed a sharp bone)?
- Has it suffered a pelvic fracture in the past?
- Does the cat show other signs of illness?
- Is it receiving any kind of medication for another condition at present?

Weight Loss

THE WEIGHT OF A NORMAL, HEALTHY ADULT cat should fall between 2.75 and 5.5kg (6 and 12lbs). Oriental Shorthairs are at the lighter end of the scale, longhairs at the heavier end; Maine Coons may even reach weights of 8kg (18lbs) or more. Be aware of what is normal for your cat's breed, type and age – check with your vet if you are not sure what this is. Also, remember that outdoor cats are likely to be lighter than more sedentary ones who live exclusively indoors.

The easiest way to calculate the weight of your cat is to hold it while you stand on the bathroom scale. Note the weight, then put the cat on the ground. Subtract your weight from the combined total – the difference is the cat's weight. You'll get only an approximate reading this way, so check it again in a week or two. If you think there's a problem, have your vet weigh the cat.

Animals quickly become thin if they go without food, and there may be an obvious reason for a sudden drop in weight. If yours is an outdoor cat, has it been missing for a day or two? Perhaps it has been locked in somewhere without food. Or has your neighbour, who feeds it treats, gone on holiday? Is there a new pet in the family? If your cat eats only a little of its food at a time and then returns to it later, another cat or dog may be eating out of the dish meanwhile.

● ***My cat Serafina spends all summer outdoors and is a skilful hunter, catching and eating mice. Nevertheless, she always seems to get thinner between June and September. Do you think this is normal?***

Have you considered that Serafina may look thinner because she has lost her thick winter coat? It's also likely that she is missing some of her regular home meals, preferring to stay outdoors. Though she is hunting more, she will get fewer calories from the mice she catches and eats, but hunting uses up a lot of energy. These are normal seasonal weight changes, and I don't think you need to worry.

● ***Bart, my 6-year-old neutered tom, appears to be losing weight. He seems a little depressed and doesn't always turn up for meals. This seems to have started after we moved. Is there a link?***

There could be – loss of appetite is a common symptom of stress, and it is likely that Bart is upset by the move to strange surroundings. Give him more time and affection than usual, but have your vet examine him in case he has an underlying problem.

● ***My 14-year-old cat, Lucky, is becoming noticeably thin, with protruding spine and hips. Yet she is eating more than usual. She has also become rather bad-tempered and hyperactive. Is it just her age?***

It seems likely that Lucky has a thyroid tumour, which is common in older cats. As a result, she is producing excessive thyroid hormone. This speeds up her metabolism and causes her to lose weight in spite of eating more. Restlessness and bad temper are common signs. She may also have increased thirst, a rapid heart rate and a dull, dry coat. The condition can be successfully treated with drugs or surgery.

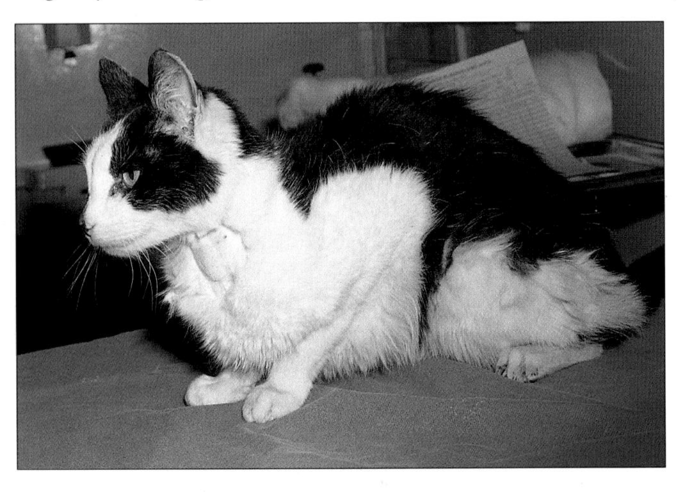

◀ *The enlarged thyroid gland can be seen as a lump in the throat of this emaciated cat, which is suffering from advanced hyperthyroidism.*

A change of diet could be the problem. A cat will simply stop eating rather than eat something it dislikes. It's no good hoping it may come to accept the change – much better to return to the previous food. Is your cat on a weight reduction programme because you thought it was too heavy? It could be lacking essential nutrients – always ask your vet's advice about diets.

Unexplained Weight Loss

You should be concerned about the cat's condition if you notice that the spine has become more prominent and the hip bones are sticking out. The face may look more angular, and the eyes seem sunken or more prominent. In a cat with a medical problem, the abdomen may sometimes increase in size due to fluid retention or an internal tumour; the cat's overall weight stays the same even though its condition is deteriorating and it is losing weight elsewhere.

Unexplained weight loss must be thoroughly investigated by a vet. One of the first things you should check for is intestinal worms (see pages 90–91), so it may be helpful to take a stool sample with you. The vet will give the cat a thorough examination, including looking inside its mouth. Dental disease resulting in painful teeth or sore gums is a frequent cause of self-starvation, especially in older cats. Infectious diseases such as cat flu or leukaemia will often depress the appetite, as will chronic pain or chronic failure of the heart, kidneys, or liver. Weight loss can also be an early indication of a growing tumour that is taking away essential nutrients.

Although the cat is losing weight, it may be eating more than usual. In this case, the vet will investigate a range of conditions that interfere with the digestion and absorption of food (see Increased Appetite, pages 112–113), as well as metabolic diseases such as diabetes or hyperthyroidism (see Increased Thirst, pages 114–115).

What To Tell Your Vet

- Have you noticed any recent alterations in your cat's eating habits? Is it leaving its food untouched, or does it seem to be eating more than usual?
- Have you made any changes to the cat's diet?
- Have you noticed any signs of worm infestation, such as rice-like segments around its bottom (see pages 90–91)?
- Does it have bad breath? Have you noticed it dribbling or having difficulty in eating?
- Has the cat had any recent bouts of vomiting or diarrhoea?
- Is the cat drinking more than usual?
- Have you felt any unusual lumps when stroking your cat?

▶ *If your cat is suffering from loss of appetite because of illness, you may have to coax it to eat by feeding it from a spoon. Your vet will prescribe a special palatable diet. It sometimes helps to heat the food to body temperature.*

Increased Appetite

It is hard to say what is normal appetite in a cat. A small, active young adult has much higher energy needs than a larger but older and lazier neutered tom, and will eat vastly more in a day. Even an individual cat's appetite varies from day to day. Some cats are very fussy eaters and will eat large amounts one day but refuse food the next. Although most owners feed their cat twice daily, many cats prefer to eat smaller quantities at more frequent intervals. Some may even visit the food bowl up to 10 times over a period of 24 hours, even eating at night.

There are times when it is normal to expect your cat's appetite to increase. Females will eat more during the later stages of pregnancy and in the first few weeks of rearing a litter. Kittens and adolescent cats need extra food for growth and increased activity. This is also true for cats that are highly active outdoors. Many cats hunt purely for the sport, preferring to bring their trophies home instead of eating them. In these cases, the extra calories are used up during the cat's physical exertions and not stored as fat.

Obesity in cats, however, is not uncommon, particularly in elderly, inactive animals. The first sign of excessive weight gain is a prominent fold of fat that runs from the flanks to the hindlegs. Very often the waistline between the last ribs and the hips disappears. If you notice signs of weight gain in your cat, check how much you are feeding it and cut down the amount if necessary. Ask your vet for advice on reducing diets.

▼ *Hunger has turned this normally well-behaved cat into a kitchen sneak thief. There may be an obvious cause, but it should certainly be investigated.*

▲ *An increase in appetite may be a warning sign of diabetes. The disease can often be managed with a daily injection of insulin, as shown here.*

Medical Reasons for Hunger

A sudden increase in appetite may be caused by a medical condition, so it should always be investigated by your vet. Kidney disease, a poor diet, digestion disorders, failure to absorb nutrients and wasting associated with cancer can all cause increased appetite, but the cause may also be routine, such as an infestation of worms (see pages 110–111). Abnormal hunger may or may not be accompanied by weight loss.

If the cat has diarrhoea (see page 107 for information on identifying this) or is passing bulky, greasy, smelly stools that contain a high proportion of undigested food, it may have a digestive disease, such as a pancreatic disorder. As a result, food is passing through the cat's system without being properly digested and absorbed, so the cat is perpetually hungry.

Sugar diabetes (diabetes mellitus) also has the effect of artificially increasing appetite because sugar – essential food for energy – is being lost in the urine. Other signs include weight loss and increased thirst (see pages 114–115). Thyroid tumours, leading to overproduction of the thyroid hormone that speeds up the body's metabolism, have a similar effect (see page 110).

Certain drugs may also cause an increase in appetite, so be sure to inform the vet if your cat has recently had any kind of medical treatment prescribed by a different practitioner.

● ***Harry has been eating and drinking more for a while and has just been diagnosed as having diabetes. I'm devastated. Will he have to be put to sleep?***

Diabetes is a failure of the pancreas to produce sufficient insulin and is not uncommon in cats. If Harry is still reasonably bright and well, there is no reason to consider euthanasia at this stage. The disease cannot be cured, but your vet will discuss with you how best you can control it. Sometimes this is done by following a strict routine of insulin injection, controlled feeding and urine testing, but if this does not suit your or your cat's habits and lifestyle, special diets and oral drugs can often be helpful.

● ***Misty has recently become extremely greedy. She devours all her regular meals and then follows me around the house begging for food. She will steal any food left lying around. Her stools are foul-smelling, pale in colour and very bulky. She's also losing weight. Should I feed her more?***

It sounds as if Misty may have a disease of the pancreas, the organ that produces enzymes to digest food. If the pancreas is not functioning properly, undigested food, particularly the fat component, becomes visible in the cat's faeces. That's what is giving her stools their particularly unpleasant appearance and smell. Feeding Misty more will only aggravate the condition, but it can be treated. See your vet, who will prescribe a special low-fat diet and an enzyme supplement for Misty to take by mouth.

Signs of Increased Appetite

If your cat's feeding habits change and you can find nothing to account for it, there may be a problem. Look for these signs:

- Your cat follows you around all day, meowing for food. It demands feeding at unusual times.
- It accepts any cat food, though it is normally a fussy eater.
- It cleans its dish in one sitting, whereas before it would eat little and often.
- It steals food from the kitchen counter, even though it has been trained not to do so.
- It begs food from you when you are at the table and pesters you for snacks.
- It visits nearby houses claiming to be hungry – check with your neighbours.

Increased Thirst

▲ *Many healthy cats dislike the taste of water from the kitchen sink and prefer to drink from outdoor sources such as ponds.*

INCREASED THIRST, ESPECIALLY IN OLDER CATS, can sometimes be a sign that all is not well, so the observant owner should look out for signs of abnormal drinking (see box). Cats were originally desert-living animals and normally have low fluid requirements due to the efficiency of their kidneys. They may be observed drinking slightly more than usual in hot weather, but they are too sensible to exert themselves and will avoid overheating by finding a cool and shady place to rest in; consequently, they do not have to drink copiously, as dogs do, to replace the fluids lost in keeping cool. There are other circumstances when a rise in drinking is to be expected, for example, in late pregnancy or while a female is feeding her kittens. Your cat's thirst will also increase if it is allowed to develop a liking for salty foods such as cooked bacon or potato crisps, or if it is begging food from neighbours or scavenging from dustbins. Don't be misled into thinking that your cat is thirsty because it is drinking lots of milk. Cats will drink water if they require fluid. Milk is not particularly good for cats, though some may develop a liking for it if it is offered to them.

There may be an obvious medical reason why the cat is drinking excessively. Acute vomiting and diarrhoea will cause extreme dehydration. Body fluids lost from wounds and burns following accidental injury will need to be replaced by drinking. Some prescribed drugs may cause an increase in thirst, and in these cases your vet should advise you to keep the cat's drinking bowl filled as part of its recovery programme.

What To Do?

If there is no reasonable explanation why your cat is drinking more than usual, you should consult your veterinarian without delay, especially if the cat is urinating excessively too. Before the visit it is very helpful to measure how much the cat is drinking. Fill and refill the water bowl from a measuring jug for 24 hours and then tip the remainder back: what is subtracted is the amount the cat has drunk. If possible, do this over two consecutive 24-hour periods to allow for variation. For best results you might consider keeping the cat inside all day. Of course, your cat may drink from other sources too, or other cats use the same water bowl, so make any necessary adjustments. Cats will often drink from a toilet, so keep the lid down. It is also sensible to take a urine sample with you for analysis. Advice on collecting urine from a cat – always a tricky operation – is given on page 116.

One of the most common causes of increased thirst in elderly cats is chronic renal failure, in which the cat passes watery urine as the kidneys cease to function: other symptoms are weight loss, bad breath and mouth ulcers. Liver disease (most common in older cats), hyperthyroidism caused by a tumour of the thyroid gland, cancer (particularly of the lymphatic system) and diabetes mellitus (sugar diabetes) are also possible causes. Urine and blood tests, X-rays, and an ultrasound scan may all form part of the diagnostic process, and the vet will then advise you on appropriate treatment.

Signs of Increased Thirst

- Your cat's water bowl needs refilling more and more often.
- The cat hangs around the water bowl and is reluctant to wander too far from it.
- The cat begins to drink from sources of water such as kitchen sinks, dripping taps, even the toilet bowl (many cats do this normally, so take note only if it is unusual in your cat).
- It drinks more than usual from puddles and garden ponds.
- The litter tray is constantly flooded.
- The cat makes more frequent trips outdoors in order to urinate.

▲ *A dripping tap provides a handy source of water for a cat suffering from excessive thirst. Take note if your cat starts to visit unusual places to drink – it may have a medical problem.*

● ***Ginger, my elderly male cat, has been drinking more and more. The vet has diagnosed chronic renal failure. Can it be cured?***

Ginger's kidneys have been damaged over his lifetime and cannot now be repaired. Transplant surgery is not yet regularly available for cats but may become so in the future. If Ginger's disease is not too advanced, rehydration by intravenous fluids (a drip) may flush toxins from his body that the kidneys have been unable to clear. Other medicines may help, and special diets are often recommended. If nothing can be done to relieve Ginger's suffering, euthanasia may be the kindest course of action. Discuss it with your vet.

● ***I've recently switched my cat Sphinx to dried food from tinned. She seems to want to drink more and has even started hanging around the dripping shower head. Is there a connection?***

Yes—tinned cat foods contain up to 87 per cent water, sufficient to meet some cats' fluid requirements. A cat fed on dried food must have water always available to make up for what is lacking in the diet. It would have been better to introduce Sphinx's change in diet gradually to give her time to adapt.

Urinary Problems

IT IS NOT UNUSUAL FOR A CAT OWNER TO SEE blood in a cat's urine or to notice it straining to pass water. Both are likely to be signs of a disease of the urinary tract, the bladder and the urethra, through which urine is passed out of the penis or vagina. Elimination of urine is essential to maintaining the chemical balance of your cat's body; a buildup of retained urine can poison the whole system, so ask your vet to investigate the problem without delay. An external examination of the cat may not reveal much, so what you can tell the vet about the cat's behaviour at home will greatly assist diagnosis (see box). Sometimes the vet can obtain a small amount of urine for testing by squeezing the cat's bladder or inserting a needle to drain it, but if the bladder is empty, it may be necessary to insert a catheter. It will help if you take a urine sample with you.

Further tests may be required. These include testing blood samples for infection and kidney function, and ultrasound and X-rays to inspect the bladder. Radiopaque dyes may be injected to show the inside of the urinary tract.

Some Common Problems

The most common problem of the bladder and urinary system in cats is urinary tract infection (UTI). Infection can ascend the urethra from the penis or vagina to the bladder, causing inflammation (also called cystitis). Since female cats have shorter urethras than males do, they

● ***Basil, my Persian cat, couldn't pass any urine, so the vet removed his penis. Was this necessary?***

The narrowest part of the male plumbing is in the penis; this is where most obstructions occur. If they cannot be flushed out any other way, the solution is to remove the penis and make a new opening for urine to come out under the tail. Although initially unpleasant, it usually solves the problem.

● ***Is an attack of cystitis an emergency? Should I rush my cat, Sophie, to the vet late at night?***

Cystitis is uncomfortable, but the cat is not really sick, and females are unlikely to develop an obstruction of the urethra. Unless Sophie is very distressed or vomiting, it is probably safe to wait until morning.

● ***How do I take a urine sample from my cat?***

Most cats will use an empty litter box if kept indoors. Wash, rinse and dry the box first, then transfer the urine to a clean bottle with a secure seal.

● ***My 11-year-old rescue cat has been diagnosed with chronic renal disease. What does this mean?***

Chronic renal disease (CRD) is common in older cats with a history of urinary tract blockage. The cat may be thin and have a poor appetite and bad breath. The cat urinates a lot, and protein may be present in the urine. Your vet will need to supervise closely and treat the condition with a special diet and medication.

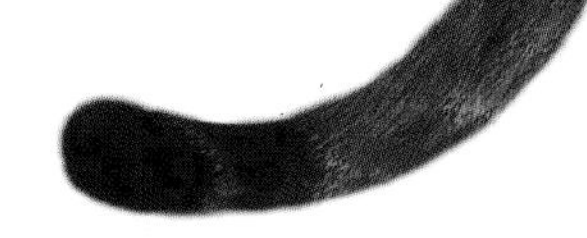

▶ *Making constant trips to the litter tray or urinating in unusual places may indicate a urinary tract infection. If it's not treated quickly, a serious problem may develop.*

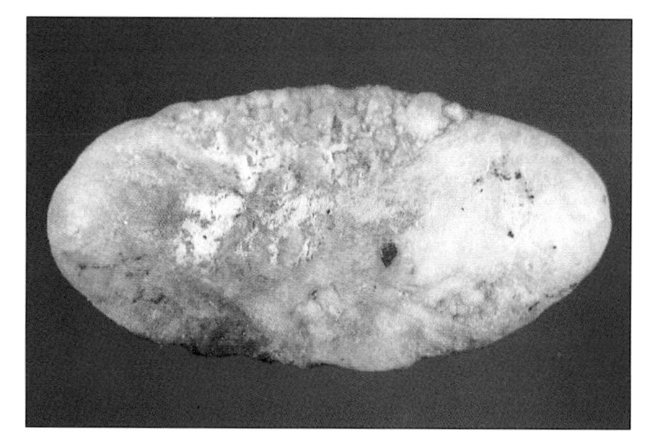

▲ *Struvite is made up of crystals of magnesium ammonium phosphate that form in cat urine. The crystals block the urethra, causing pain and acute infection. Sometimes stones are formed, as here. Rarely they may grow up to 13mm (½in) long.*

What To Tell Your Vet

- Is there blood or streaks of blood in the urine?
- Is the cat urinating more frequently than usual, and producing only small amounts?
- Have you noticed spots of urine or blood around the home?
- Is the cat urinating in unusual places?
- Is it repeatedly squatting and straining but failing to urinate? Does it seem distressed or in discomfort?
- Is it licking its genital area? Have you noticed blood on the genitals? Is the penis extruded a lot of the time?
- Is it drinking more than usual?
- Is it lethargic or lacking in appetite? Has it collapsed at any time?

are more susceptible than males to UTI. The inflamed bladder wall is irritable, so the cat will strain even when its bladder is empty. The bladder wall may also bleed. If left untreated, bacterial infections can spread from the bladder wall up the ureters, the tubes that pass urine from the kidneys to the bladder, and eventually reach the kidneys. This can be life-threatening.

Antibiotics will usually clear up most simple infections, so early treatment is always advised. Prolonging treatment reduces the likelihood of recurrence, so be sure to complete the course of medicine prescribed by the vet, even if the cat's symptoms have subsided. You should encourage the cat to drink as much as possible, as this helps flush the infection out of the urinary tract. One way to do this is to add a little salt to your cat's food. Your vet will advise you.

Other urinary problems include feline lower urinary tract disease (FLUTD), also known as feline urological syndrome (FUS), which mostly affects males. It has many different causes, both bacterial and viral, and may be associated with the presence of crystals in an alkaline urine. The urine becomes concentrated, and the salts in it form a sandy deposit of crystals or tiny stones that block the urethra. This is particularly narrow in males, and if it becomes entirely blocked with debris, the cat can no longer pass urine, causing a medical emergency. Polyps or tumours in the neck of the bladder may give the cat pain or discomfort in urinating. Tumours often ulcerate and bleed, so there are likely to be streaks of blood in the cat's urine. A ruptured bladder following a street accident or a hard fall may also cause major problems if not diagnosed.

Partial or complete obstructions of the urinary tract must be relieved as quickly as possible. In some cases they can be flushed away; others may require surgical intervention. Special diets sometimes form part of a treatment programme. Polyps can be removed, but some tumours of the urinary tract may prove to be untreatable.

Preventing Urinary Diseases

It is extremely important to make sure that your cat always has access to fresh water, especially if you are feeding it only dry food. Though many cats prefer to drink from puddles outside or dripping taps, you should always keep your pet's water bowl filled (see Increased Thirst, pages 114–115). Some formulated cat foods are low in magnesium, and this may help produce an acid urine, which reduces the likelihood of crystals forming. Ask your vet for advice.

Cystitis is more common in the winter among outdoor cats, who may become reluctant to go outside to urinate because of the weather. As a result, they avoid going out, and the bladder becomes overfull. Make sure your cat has access to a litter box, or see that it goes outside frequently to urinate in a suitable area.

Sneezing

ALL CATS SNEEZE FROM TIME TO TIME, JUST AS humans do, as a reflex reaction to an irritation of the nasal passages. Inhaled substances such as talc or dust (if the cat has been crawling under the bed, for example) will often cause a sneezing fit, and some cats may develop a sensitivity to aerosol sprays, but allergies are extremely rare. A cat that is suffering frequent or severe outbreaks of sneezing is much more likely to have a mild or severe form of cat flu (see pages 86–87), especially if there is any sort of discharge from the nose and eyes (one that is clear and watery is generally a better sign than one that is thick, yellowish green in colour, or bloodstained). The cat may also stop eating and appear listless.

You should see the vet as soon as possible, especially if your cat's vaccinations are not up to date and it has been exposed to other felines. Call the vet first. As cat flu is highly contagious and there may be unvaccinated kittens waiting in the reception area, leave your kitten or cat in the car while you go inside to inform the receptionist of your arrival.

Meanwhile, keep the affected cat indoors. You can relieve its symptoms by bathing the nose and eyes gently with warm water. You can try tempting its appetite by serving small amounts of warm, aromatic food. Do not give your cat any human flu remedies that can be bought over the counter – most of them are likely to be extremely dangerous to cats.

Kittens are at particular risk from cat flu, especially if their mother is a carrier of one of the flu viruses. The virus is shed into the environment in the queen's saliva, urine, or faeces, exposing the kittens before they are old enough for routine vaccinations. Cat flu can be treated successfully in most kittens but they run the risk of becoming carriers of the virus and prone to recurrent bouts of flu or a stuffy nose throughout their lives.

◀ *A kitten gets that familiar tickly feeling (left) and puts his paw up to his mouth as air is expelled from his nose (below).*

Other Causes of Sneezing

If your cat has suddenly started sneezing but does not otherwise appear to be ill, and especially if only one nostril appears to be discharging, it may have a nasal obstruction. A possible cause is swelling from an abscessed tooth at the top of the mouth. Sometimes, after a cat has been eating grass, a blade of grass becomes caught at the back of the throat and, instead of passing into the oesophagus (food pipe), is propelled forwards into the nasal cavity. You may sometimes be able to see the tip of the grass blade actually protruding from the nostril. The cat usually succeeds in sneezing it out. Don't ever be tempted to poke an implement up the nostril to dislodge an obstruction – this is likely to damage the delicate membrane lining. It is much better to have the vet deal with the problem.

A discharge from the nose that is thick, yellow, or green in colour, or bloodstained, always calls for speedy veterinary attention. Fungal infections or tumours of the nose are among the possible causes your vet will want to investigate, almost certainly by taking an X-ray and possibly a biopsy. Only then can the appropriate course of treatment be decided.

▼ *Having sneezed, he wipes his nose and face dry with a delicate paw.*

▲ *Severe irritation of the nose has caused heavy sneezing in this silver tabby longhair, with a blood-stained discharge. Until the cause has been determined, the cat should be isolated from other felines.*

● ***Minnie, my tabby, has been sneezing intermittently for a while, but I have not seen any discharge from her nose and she seems quite well. I've now noticed a small swelling below her right eye. Could they be linked?***

Yes. These symptoms suggest that an abscess on the root of a tooth is causing swelling and irritating the sensitive nasal passage. Open her mouth gently and examine the teeth and gums on the upper righthand side for signs of tartar, redness, or pus on the gum. Her breath may smell offensive. Your vet should examine her, probably under general anaesthetic. If a tooth abscess is ruled out, an X-ray and biopsy may be necessary to establish the cause of the swelling.

● ***A while ago we took in a stray tom. We had him neutered and vaccinated, and he has settled well. However, he is sometimes sneezy, with a creamy discharge. Could this be a reaction to the flu shot?***

Your tom may be a carrier of a type of flu virus due to exposure in his earlier roaming days. The stress of neutering may have caused the symptoms to reappear. The vaccine will protect him against other types of flu, but he may get recurrent bouts of the sniffles throughout his life. If the discharge doesn't clear up, or gets heavier, the vet may prescribe antibiotics and have him tested for feline leukaemia and FIV (see pages 86–87). Meanwhile, keep him away from unvaccinated cats.

Respiratory Problems

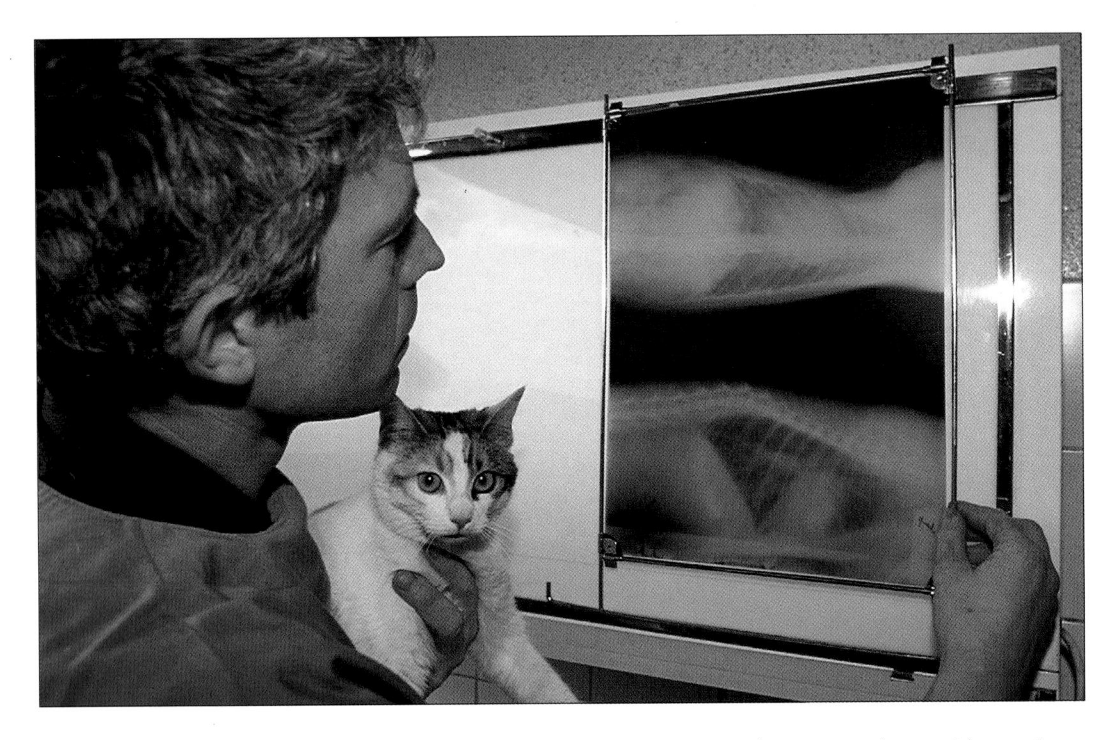

▲ *A vet examines chest X-rays of a cat with a persistent cough. It is usually necessary to sedate or anaesthetise the cat in order to take X-rays.*

COUGHING DOES NOT OCCUR VERY FREQUENTLY in cats. The cause of the problem normally lies within the lungs or heart, or (less often) in the trachea, the airway from the mouth to the lungs. An occasional dry cough is usually no cause for concern, but any cough that persists and increases in frequency or severity should be investigated by your vet. It will help if you can describe the cough: when you first noticed it, whether it is dry or moist, how often it occurs, whether it is more obvious when the cat is moving or at rest. The vet will also want to know if the cat is up to date with its cat flu vaccinations (see pages 86–87), if it shows other signs of ill health such as loss of appetite, and if other cats in the household are affected. It is likely that an X-ray of the cat's chest will be taken. The vet may also view the back of the cat's throat and the airway with an endoscope to check for any blockages or other abnormalities, and take a sample of mucus for bacteriology or microscopic examination.

Difficulties in Breathing

A cat at rest breathes normally and soundlessly through its nose at a rate of about 25–30 breaths a minute. Normal purring has no connection with breathing – it is generally a sign of pleasure – but occasionally a louder, deeper purr is heard when a cat is in pain or ill. Other sounds associated with your cat's breathing may indicate a problem. Open-mouthed breathing, rare in cats, usually occurs when the cat is distressed or in pain, or may be a sign of infection or underlying heart or lung disease. Check to see if the cat's nostrils are blocked with mucus, blood, or pus, and clean them gently with moist cotton wool. Do not poke anything up the nostrils.

A cat with extreme difficulty in breathing (dyspnea) tends to crouch, extend its neck forwards, push its elbows away from its chest wall, and use its abdominal muscles to help push air in and out of the lungs (you will see its tummy muscles visibly moving). Handle a cat in this state extremely carefully as any extra stress or panic will make breathing even more difficult. If your cat appears to stop breathing altogether, lay it flat on its side, pull its tongue gently forwards to clear the air passage to the throat, and check that nothing is stuck at the back of the throat (take care not to get bitten). Sometimes pulling the cat's tongue gently can stimulate breathing (see First Aid and Emergencies, pages 134–137).

Get the cat to a vet as quickly as possible. It will help if you can give precise and clear information about the attack. Was it sudden, or did it occur gradually over a period of time? Did it come on when the cat was moving about? Has the cat been in an accident or fight? Does it seem to have pain elsewhere in its body? Are there any other signs of respiratory disease such as runny eyes and nose, a cough, or poor appetite?

It is very likely that an X-ray of the chest will be required. As cats with breathing difficulties present a higher than usual anaesthetic risk, initial investigations are often carried out under sedation only. However, sometimes a surgical repair may be urgently needed – for example, if the diaphragm has been ruptured in a street accident – and then an anaesthetic will be needed.

● ***My 2-year-old cat, Gemma, has bouts of open-mouth breathing in the spring, and her tongue turns slightly blue. The attacks only last a short time. What could they be?***

Gemma may be very sensitive to a seasonal allergen such as pollen, causing a brief asthma attack (a spasm of the airways). So far this hasn't distressed Gemma too much, but it may get worse each year. A course of steroids each spring may prevent or reduce the attacks.

● ***My 18-month-old tomcat, Elmo, has had a cough ever since I can remember. My vet examined him at 6 months and said his chest sounded clear. He has grown well, his eyes are bright, and he is outside hunting all day. Should I worry about his cough?***

As Elmo has had the cough for so long and it doesn't seem to be distressing him, it is difficult to justify putting him through extensive tests. However, as Elmo is a hunting cat, I would advise asking your vet to test him for lungworm (see pages 90–91).

● ***My elderly tabby, Zoe, has lost weight recently and developed a dry cough. I know she has some small lumps under her skin on her tummy, but they don't seem painful to touch. What should I do?***

Zoe could have cancer of the mammary glands. It spreads quickly, often to the lungs, so you must have the lumps investigated as soon as possible for Zoe's sake. If they turn out to be malignant, the vet may advise a chest X-ray; following this, he or she will discuss palliative treatments with you.

Some Causes of Respiratory Problems

Signs	Possible cause	Action
Sudden-onset cough. Watery eyes and nose, sneezing	Cat flu, inhaled irritant, e.g., aerosol spray	Check if up to date with flu shots. See vet
Bout of sudden-onset coughing; sneezing	Exposure to irritants	Remove cat from area
Dull cough, cat not eating and depressed, severe dyspnea	Pneumonia or serious lung disease, fluid or pus in chest, chest tumour	Keep cat as relaxed as possible. Transport quickly and gently to vet without delay
Open-mouthed breathing, lips and tongue look blue	Asthma	Keep cat calm in well-ventilated room. See vet
Open-mouthed breathing with 60–90 breaths per minute	Heatstroke, chest injury, poisoning, nasal obstruction, cardiac disease	Cool cat with tepid water (see pages 134–137). Check for any sign of injury or access to poisons. Clean blocked nostrils. Take to vet immediately
Noisy breathing (snorting, snuffles, or choking)	Foreign body, mass, or tumour obstructing back of throat or nostrils	Take to vet. You can remove a foreign body at the back of the mouth with tweezers (see pages 134–137), but do this only in an emergency

Collapse

SOMETIMES CATS COLLAPSE WITH NO WARNING or obvious explanation – the owner finds the cat limp and motionless, unaware of what is going on around it. It is visibly breathing, but the rate may be either faster or slower than normal (see pages 120–121). Try to make a mental note of everything that is happening – or better still, if you have a chance, write down all the details you remember about the circumstances of the collapse. What you can tell the vet later will be of vital importance (see box).

If you find your cat collapsed, it is extremely important that you stay calm. Move the cat very gently out of immediate danger, for example if it is lying in the street. Ensure that nothing is obstructing its throat or nose, and turn it onto its side with its neck gently stretched out and its head slightly lower than its body. Pull the tongue out and to one side. Check the cat's breathing and heart rate, and give it artificial respiration if needed (see First Aid and Emergencies, pages 134–137). Keep the cat warm and call the vet for advice without delay. Do not give any medication or alcoholic beverages such as brandy.

A sudden collapse can be caused by shock. This can be the result of internal haemorrhaging following a street accident or a bad fall, internal rupture, or acute metabolic failure. In shock, the body shuts off the blood supply to the extremities such as ears, feet and tail, and concentrates on keeping oxygen flowing steadily to the heart and brain. The cat's extremities feel cold, and it seems barely conscious. Without rapid treatment, shock can be fatal, especially in an elderly cat or one that is recovering from a severe illness. Rush it to the vet for emergency treatment.

What To Tell Your Vet

- How long did the collapse last?
- What was the cat doing just before it collapsed? Was it waking up or exerting itself?
- Has the cat recently been injured in a fall, fight, or street accident?
- Have you noticed it behaving oddly over the last few days or weeks?
- Is it currently taking any medication?
- Has the cat had any previous collapses? If so, when? How long did they last?
- How long did the cat take to return to normal?
- Were there any after effects? Did the cat seem subdued or dazed?
- Was it hungry or thirsty on recovery?

◀ *If a cat collapses from shock, it must be kept warm. Lay it on its right side and wrap it gently in your coat or sweater. Seek immediate help from a vet.*

Why Cats Faint

Temporary collapse, or fainting, occurs when the body is starved of a vital factor such as oxygen or glucose. For example, if the supply of oxygen to the brain is disrupted because of a circulatory problem such as heart failure, or by the obstruction of an airway, the body's systems simply close down and the animal collapses. This makes it easier for the heart to pump oxygen directly to the brain again, and the cat can recover after 1–5 minutes. Unless the cause is serious, after effects should be short-lived. The cat may collapse only once, but collapses can sometimes occur repeatedly. The intervals between episodes may be as much as 2 or 3 months, but it is not unusual for them to take place more frequently.

The briefness of a collapse can make it difficult to identify an underlying problem. Blood tests will measure such things as glucose levels, but the results may not be reliable unless the sample was taken while the cat was collapsed. Similarly, a heart condition that is causing occasional collapses may not be detectable simply by listening to the heart and lungs with a stethoscope once the cat has recovered. If the attacks are occurring quite frequently, the cat may have to be hospitalised so that its reactions and bodily functions can be observed and tested while it is actually experiencing an attack. Only then can a reliable diagnosis be made and treatment started.

● ***Our 3-year-old Maine Coon, Freddie, had a runny nose for a couple of days; then he started to flop down every time he walked a little way. He isn't in pain and can eat, but his 'meow' changed too. The vet thinks he may have myasthenia gravis. What is this?***

Myasthenia gravis is a rare neuromuscular condition that would fit the symptoms you describe. It can usually be treated with steroids. Your vet will probably want to do some blood tests to establish the diagnosis, and will then prescribe medication.

● ***About two years ago, Moonchild, my 8-year-old cat, started to have brief collapses. They occur about 4 times a year. The vet hasn't been able to find anything wrong, and a recent blood test was normal. I'm really concerned about it. Is there anything else I can do?***

It can be difficult to make a diagnosis if the collapses are too short for your vet to see or test the cat when they are occurring, especially if the cat appears to be absolutely normal between collapses. You could leave Moonchild at the veterinary clinic so that she can be observed and tested during a collapse, but this is not very practical if they are happening so infrequently. Why don't you try to record one of the collapses on a camcorder? Then your vet will be able to see what happens when Moonchild collapses and have a better idea of what is causing them.

Some Causes of Collapse

SIGNS	POSSIBLE CAUSE	ACTION
Limp body, cold extremities (e.g. ears, feet, tail), pale gums, rapid or slower breathing	Shock	Keep warm, rush to vet without delay
Limp body, blood from nose or elsewhere, torn nails	Street accident	Keep warm, rush to vet without delay
Limp body, close to electricity supply	Electric shock	Do not touch cat until electricity supply is switched off. See vet immediately
Limp body, rapid breathing, extremities (ears, feet, tail) very hot	Heatstroke, if cat has been confined in hot car or room	Bring down body temperature with tepid or cool water, but beware overcooling. See vet even if cat recovers quickly
Cat goes limp but recovers quickly	Fainting due to lack of oxygen to the brain. May occur after exertion or prolonged bout of coughing	Make cat comfortable. Give reassurance on recovery. See vet

Seizures and Convulsions

MOST KITTENS TWITCH WHEN THEY ARE asleep. Sleeping cats, too, often make sporadic restless movements as if dreaming. These movements are very different from what happens when a cat has a seizure (also called a fit). Seizures are caused by a disturbance of the brain that results in involuntary and uncontrollable activity of the muscles. The cat may show only mild behavioural changes such as repetitive biting at imaginary objects, or more severe locomotive disorders such as paddling or running movements with the legs, excessive arching of the neck and spine, and possibly loss of consciousness. At the same time, the cat may salivate excessively and lose control of its bladder and bowels. Though a seizure can be very frightening and distressing for an owner to witness, in most cases the cat is not experiencing any pain and is unaware of what is happening.

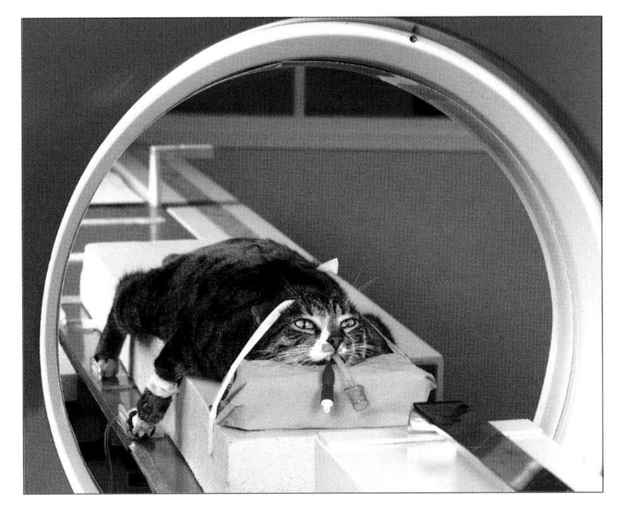

▲ *In some cases of neurological disorder it may be necessary to carry out a CT scan of the brain. The cat must be kept absolutely still and will be anaesthetised for this procedure.*

If your cat is having a seizure, it is best not to intervene, however much you may wish to comfort it. Do not try to move the cat, but prevent it from damaging itself. Keep quiet, shut off all lights and shut off the radio or TV. If outdoors, shade the cat from glare. Remove sharp or hard objects from the immediate vicinity of the cat. Try to observe exactly what is happening and time how long the seizure lasts. Writing down this information will help you give an accurate description to the vet later (see box).

The normal duration of a seizure is 1–3 minutes. Leave the cat to recover quietly and take it to the vet for an examination several hours later. However, if the fit lasts longer than 10 minutes, you should call the vet immediately for advice.

Characteristic Signs of a Seizure

STAGE 1: AURA

Early onset

- The cat seems disoriented or dazed
- It appears wobbly when walking
- It may become aggressive
- It may become very hungry
- It may start calling and crying more than usual

STAGE 2: ICTUS

The fit itself

- The body becomes rigid ('tonic' phase)
- The cat has muscle spasms and tremors ('clonic' phase) It makes paddling movements with its legs and the neck arches backwards
- The cat may lose consciousness
- The cat may urinate and defecate, and salivate profusely

STAGE 3: POST-ICTUS

After the fit

- The cat is confused and seeks reassurance
- It is hungry and thirsty
- It may appear disoriented for a little while afterwards

• *My 16-year-old male cat, Archie, has started to behave oddly. He growls at the wall, eats ravenously and paces about. He has had three short fits over the last month when he falls to one side, paddles his legs and loses control of his bladder. Could he have a brain tumour?*

A brain tumour could cause the signs you describe, but so can a number of other problems. You must take Archie to the vet – simple blood tests will establish whether kidney or thyroid disease, common in elderly cats, are responsible. Neurological problems such as a brain tumour can be very difficult to diagnose, and further tests usually require referral to a specialist. Your vet will discuss fully with you all the options for investigation and treatment that it might be reasonable to consider for a cat of Archie's age.

• *About 6 months ago Pandora, my 6-year-old spayed Siamese, was diagnosed as having epilepsy. She now has a tablet at prescribed times twice a day, and I am sure she swallows them. However, she still has a mild fit about once every 2 weeks, though she is extremely well and happy the rest of the time. Is the medication working?*

It is possible that not enough of the drug is being absorbed into the blood so that the levels in the brain are not high enough. Your vet can check this by taking a blood sample in the interval between the times Pandora takes her tablets. A dosage change may be required, but it can take 10 days or so for this to take effect. However, some cases of epilepsy are not entirely controllable with medication, and you may need to discuss with your vet what you consider an acceptable number of fits. Other treatment options such as acupuncture may also be of help.

• *My 14-week-old Persian kitten, Shula, has started to dribble saliva excessively and throw occasional fits. They always seem to happen about an hour after she has eaten. She loves her food but does not seem to be growing or gaining weight. I have treated her for worms, but it has not made any difference. Are these just normal growing problems, or should I be more concerned?*

It is important that you take Shula to your vet as soon as possible for a thorough examination. Fits such as this in a young kitten could be a sign of a severe congenital abnormality, such as a liver defect, and the sooner you get to the bottom of the problem, the better it will be for all concerned.

What to Tell Your Vet

- How long ago did you notice the first seizure?
- How often do they occur, and how long do they tend to last?
- Does the cat's behaviour change just before the seizure? Does it become aggressive or seem disoriented and confused?
- Does it lose consciousness during the fit?
- Has its diet or lifestyle altered recently?
- Has it recently been treated for fleas or been dewormed?
- Could it have had access to poisons such as rodent or slug bait?
- Has it been hit by a car, fallen off a wall, or been bitten?

What Causes Seizures?

Any number of things can cause a seizure. The cat may have been poisoned, perhaps by slug or rodent bait. Some antiflea and deworming drugs may also cause seizures if they are used incorrectly. The cat may have injured its head in a street accident or bad fall. Seizures are also caused by abnormal levels of calcium or glucose in the blood (diabetes mellitus), or by kidney or liver failure or heart disease. Fits in a young kitten should always be investigated, as they can be a sign of some forms of congenital abnormality and may not be treatable.

If the seizures occur repeatedly over a period of time, the cat may be diagnosed as having epilepsy. This condition is much rarer in cats than it is in dogs. The severity, frequency and duration of the seizures may or may not increase with time. They are most likely to be a sign of a disorder within the brain itself, but other causes such as a liver or kidney disorder or exposure to toxins will have to be considered and ruled out first. The cat may be referred for special tests if your vet suspects it has a brain tumour or another neurological disorder. Treatment for epilepsy is not usually given until other possible causes have been ruled out as far as possible.

Although a diagnosis of epilepsy often sounds alarming to an owner, cats with true epilepsy of unknown cause are often able to live out a normal lifespan with few complications, provided that the right levels of medication are found.

Loss of Balance

▲ *A healthy cat should be sure-footed, negotiating difficult paths with ease. If your cat seems unsteady on its feet in any way, there may be a problem.*

CATS MOVE WITH NATURAL ATHLETIC GRACE. It is abnormal for them to be wobbly on their legs unless they are very old, recovering from an anaesthetic, or have a severely debilitating illness. If your cat is relatively young and healthy but seems to be unsteady on its feet, watch it carefully – it may be an early sign of a health problem. Make a point of seeing if your cat is having difficulty jumping or lands heavily. Is it beginning to use walls to aid its balance? If the cat is very unsteady, keep the cat indoors in an area where it cannot hurt itself by jumping on or off furniture or shelves, and take it for a medical examination as soon as possible. The vet will want to know your cat's vaccination history, when the loss of balance began, how quickly it progressed, whether it is worse at one time of day, whether the cat seems to be in pain, if it has been in a fight or an accident, and whether the cat shows any other sign of illness. If the cat has a harness, take it with you; watching the cat walk will help the vet make a diagnosis.

A Wide Range of Problems

Balance is controlled by the organs of the inner ear (the semicircular canals, utricle and saccule), which tell which way is up and detect movements of the head, sending this information via the vestibular nerve to the cerebellum, the part of the brain that coordinates posture and movement. Any disturbance to the ears or brain as a result of disease or accidental injury may disturb the mechanism of balance. Ear infections commonly cause an unsteady gait and balance loss,

Possible Causes of Imbalance

SIGNS	POSSIBLE CAUSE	ACTION
Unsteady on legs, head tilts to one side, jerky eye movements	Ear tumour or infection Vestibular syndrome	Check ear canals for wax/pus and clean with damp cotton wool. Consult vet without delay
Unsteady on legs, confused behaviour, dementia, loss of vision or balance	Brain tumour or infection Head injury	Take to vet as soon as possible
Bouts of unsteadiness	Metabolic disorder Mild epilepsy	Consult vet
Staggering, vomiting, dilated pupils	Poisoning	Try to identify source of poison and take cat to vet immediately
Unsteady on legs, chin tucked in, poor appetite	Vitamin B1 or potassium deficiency (esp. in older cat)	Consult your vet as soon as possible about the cat's diet

and are usually treatable with a course of antibiotics. A condition known as vestibular syndrome causes the cat to lean to one side and may be accompanied by jerky eye movements. The cat is often left with a mild head tilt.

Brain injuries, tumours, poisoning and mild epilepsy may all impair balance and movement to a greater or lesser degree; accompanying signs are sudden collapse, seizures, and depression. Another possible cause of unsteadiness and balance loss is a vitamin or potassium deficiency disorder that can interfere with the normal functioning of the brain. With this wide range of possible causes, it is essential that you have your veterinarian examine the cat as soon as possible so that diagnosis can be made promptly and the appropriate treatment started.

● ***I recently treated our new kitten with a flea powder from the pet shop, and he seemed very unsteady on his legs for a few hours afterwards. Should I throw the powder away and try something else?***

Yes. Your kitten may be sensitive to the active insecticide, or the dose may have been too great for his size. Insecticides often work by attacking the insects' nervous system and can affect the kitten's nervous system too. Make sure any product you use is labeled 'for cats' or 'for cats and dogs'. Cats' livers cannot cope with some substances that dogs' can.

● ***I've heard that raw fish will give a cat brain damage. Is this true?***

Raw fish contains an enzyme that destroys Vitamin B1(thiamine), essential for normal brain function, before the cat's gut is able to absorb it. Overcooking food can also destroy Vitamin B1. A balanced commercial diet is best for your cat, though you can give it an occasional treat of cooked fish.

● ***What causes vestibular syndrome, and how serious is it?***

Vestibular syndrome is a sudden-onset condition affecting adult cats, usually during the summer months. The cause is unknown, but most cats recover completely in 2–4 weeks without treatment. A minority may be left with a mild head tilt.

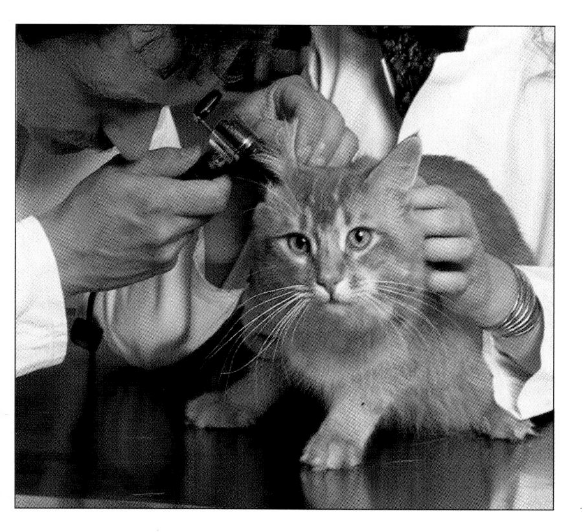

◀ *In cases of loss of balance, one of the first things the vet will do is examine the cat's ears for signs of infection. Its eye movements will also be tested.*

Lameness

IT MAY NOT ALWAYS BE IMMEDIATELY OBVIOUS that your cat is lame – many cats simply hide away and avoid moving rather than put weight on an injured leg. If you do notice that your cat is limping or hopping, take great care when picking it up or examining it: however docile your pet is normally, it may suddenly turn on you and scratch and bite if hurt. Wrap a thick towel or blanket around the cat, leaving only the injured leg exposed. Feel very gently over the affected limb, starting from the area of least pain. Look for heat, tenderness, swelling of the leg or paw, broken nails and signs of injury such as matted hair, hair loss, puncture marks and bleeding. Fraying of the claws can be a sign that the cat has been involved in a street accident.

Sometimes lameness stops after a short time and the cat is none the worse for its injury. If it doesn't, or if it recurs, take the cat to the vet. Keep a note of which leg is affected and observe the way the cat is walking. Once a cat is in the examining room, it will often become apprehensive and refuse to move, making it difficult for the vet to observe the lameness. Do not feed the cat before seeing the vet – it is likely that the vet

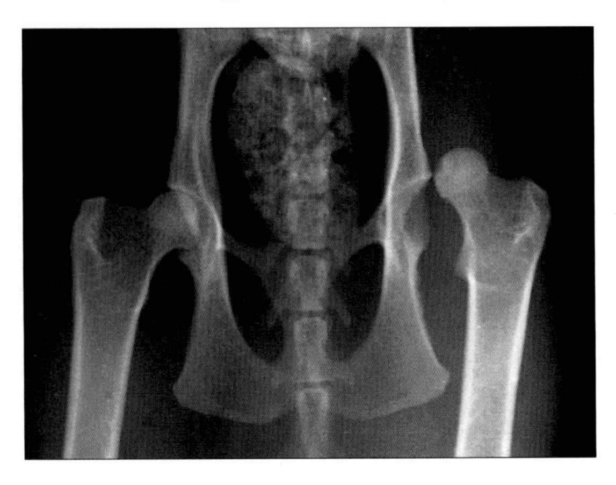

▲ *A severe dislocation of the left hip is clearly visible in this X-ray of an adult cat. Unlike dogs, cats rarely suffer from congenital hip deformities. Traffic accidents and falls from high buildings account for most problems.*

● ***After Fernando, my tomcat, was hit by a car, the vet had to amputate his leg. Fernando used to be a real tree climber. How will he manage now?***

Cats seem to have no difficulty at all in getting around on three legs, which is why amputation is always an option in cases of severe leg injury. Don't worry – I'm sure you'll find Fernando is soon climbing trees again.

● ***My last cat, Cheng, suddenly went lame and collapsed. The vet said he had a blood clot, and Cheng was put to sleep. I now have a new cat. How can I prevent the same thing happening to him?***

Cheng had a thrombosis, a condition associated with heart disease. A blood clot becomes lodged in one of the arteries supplying blood to the hindlegs, and the cat loses the use of its legs. The chance of recovery from thrombosis is poor, and most feline patients have to be euthanised. The best prevention is to make sure your new cat is fed a good-quality diet throughout its life and has adequate access to exercise.

● ***Simba, my 14-year-old shorthair, has recently started to limp. I've noticed that his claws are always slightly extended. Are the two connected?***

Are you regularly clipping Simba's claws? It sounds as if they may have been allowed to become too long and are growing back into the pad, causing him pain when he puts weight on his feet. He may also have a problem with his retracting tendons, which is why the claws are extended. Your vet should examine him as soon as possible.

● ***After our cat, Bonnie, was hit by a car she had a metal plate and screws put in her leg to repair a fracture. It seemed as if she was starting to heal, but now the leg looks swollen and she has begun limping. What could be causing this?***

You must take Bonnie back to the vet without delay. Most modern surgical devices are designed so that the metal plate can be left permanently in place as the bone heals, but there is a possibility that the screws have come loose, and this has caused the affected leg to swell. If so, Bonnie may need to have more surgery to remove the implants.

▲ *Great care is needed when examining an injured leg. Don't attempt to do so yourself if the cat seems anxious or disturbed, or if there is obvious major injury – it is better to leave this to the vet.*

will need to sedate or anaesthetise the cat in order to manipulate the leg and examine it thoroughly, and may decide to take X-rays.

The most frequent causes of lameness in cats are injuries from street accidents (see pages 132–133), a fall from a high window ledge or balcony, or an infected bite wound (see pages 130–131). Because cats are so light on their feet, cuts to their paws are relatively infrequent. However, if a cat steps on glass or sharp metal, it can result in heavy bleeding (see First Aid and Emergencies, page 136). The pads of the feet may become skinned and sore if the cat has been walking over rough surfaces or is an inveterate climber. Minor grazes can be treated at home by bathing in salted water. A surprisingly common injury to the pads are burns. This can happen if the cat jumps onto the stove at home. Place the paw or paws immediately under cold running water and take the cat to the vet. Lameness can also be caused by damaged and infected nailbeds and fractured claws.

Bone and Joint Disease

Cats tend to suffer from fewer problems of the bones and joints than dogs, though hip dysplasia (an abnormality of the hip that causes stiffness and a hopping gait) is seen in some breeds such as the Maine Coon. Siamese cats have a greater tendency than other breeds to bone malformations. Arthritis (inflammation of a joint) is seen most frequently in older cats, especially if the joint has had a previous injury. The cat will be stiff in the joint, especially on waking. You can help by massaging the area gently and providing extra padding for the cat to sleep on. Young cats and kittens occasionally suffer from an infectious arthritis, which is accompanied by a high temperature. The cat appears sick and moves stiffly. Arthritis is also caused by Lyme disease, transmitted by the deer tick. The cat shows signs of lameness about a month after infection.

Lameness may also be a sign of bone diseases associated with malnutrition and kidney disorders. The thin, weak bones may fracture spontaneously. Diseases of the nerves and circulatory system, such as a thrombosis of the iliac artery supplying blood to the hindlegs, can cause lameness and even paralysis. The cat's legs will feel cold and limp. It is intensely painful – keep the cat warm and get it to the vet without delay.

Injuries from Fighting

CATS FREQUENTLY FIGHT OVER A TERRITORY (since they do not respect the boundaries of your fence), over food or the contents of a dustbin, and sometimes seemingly just for the sake of it. If any two cats in a neighbourhood dislike each other, the fur may fly quite regularly.

The cat most likely to be involved in a fight is the intact (nonneutered) tom, who has a social position to maintain in the neighborhood hierarchy. Neutered toms are less prone to fight, unless they have been neutered after they are well into adulthood and are already used to fighting (see pages 74–75). Neutering works gradually to lower the hormone levels in the tom's bloodstream, so even a young male cat that has been recently neutered will not immediately lose the urge to fight. Nor is it unheard of for females to become involved in a fight, especially if they are defending their kittens; a mother whose litter is threatened will not hesitate to take on the toughest tom. All cats with access to the outside world

▼ *This cat has the telltale scratched face of a fighter. Scratches and surface wounds can be treated at home; deep punctures, or any injury to the eyes, should be attended to immediately by your vet.*

are potential fighters or potential victims. The only guarantee that your cat will have a peaceful life is to keep it indoors, but many people consider this an unacceptable restriction. Vaccinating your cat against feline viral diseases (see pages 86–87) will at least protect it from infection by strays or unvaccinated cats.

Fighting Tooth and Nail

The phrase 'fighting tooth and nail' applies perfectly to the typical cat fight. Cats' teeth – particularly the canines (or fangs) – are sharp, and the jaw muscles are very powerful. The pain of a bite wound can be increased by infection. A cat's mouth often harbours particularly nasty bacteria, and a bite can deposit these organisms deep into the damaged tissue.

The kind of injury that results from a fight depends to a certain extent on where the injury occurs on the body. A bite on the paws, limbs, or body of the tail typically leads to cellulitis – a hot, painful swelling of the tissues adjacent to the puncture wounds, which can sometimes track some distance away from the original site. If the bite is to the face, rump, or the base of the tail, an abscess – an accumulation of pus under the skin leading to swelling – is more common (see pages 96–97).

The nail or claw part of the cat's armoury is less likely to cause any severe damage. Most scratches are relatively superficial and will cause the cat only a little discomfort. Occasionally, however, a lashing paw can inflict a serious injury if it hits the eye. A scratched eyeball or punctured cornea may leave permanent scars, possibly leading to loss of sight in that eye or even loss of the eye itself. A cat with any eye injury should be taken to the vet immediately.

It is always a good idea to take your cat to the vet for any suspected battle injury that involves a bite. However, for a light scratch you can try home treatment by washing the wound twice daily with a diluted saline solution and keeping a close eye on it for 2–3 days to see if it swells or develops infection. If the wound heals, a visit to the vet may not be necessary. Puncture wounds are most likely to become infected and may need antibiotics. Infection can result in an abscess, which will need to be lanced by a vet.

Signs of a Fight Injury

Be on the alert if you know from experience that your cat fights regularly or if you have overheard a fight outside your house in the past day or two. You should watch out for:

- Sudden lameness in one or more legs.
- General listlessness and lack of energy.
- High temperature or hot, dry nose.
- Growling or hissing when handled.
- Visible wounds, although a single puncture wound can be very hard to locate, either by sight or touch.
- Swelling, particularly around the face, the rump, or the paws.
- Torn, bitten, or scratched ear.
- Closed, swollen, or damaged eye.

● ***Claudius, my 2-year-old male Rex, keeps getting into fights and developing abscesses. How should I treat them?***

Any abscess that Claudius develops must be lanced by your vet. Abscesses contain infected pus walled off from the rest of the body. The wall helps to protect the body but also prevents drugs from getting into the abscess and eliminating the infection. Recurrence is common if the infected material is not drained off by lancing and flushing. Also, the toxins in the abscess are more likely to make your cat ill if they remain.

● ***My tomcat, Bear, came in limping yesterday, but wouldn't let me examine his feet. Now he has a temperature, but I can't find any sign of a wound. What should I do?***

Try again to examine Bear – be careful, as you do not want to end up getting bitten! If he lets you, you will probably find a plug of matted hair covering a tooth-sized hole on one paw or leg. Even if not, you should still take him to the vet; if he has a temperature, he may have already developed an infection from the unseen wound and may need antibiotics.

● ***Can I use antiseptics from my own first-aid kit to clean my cat's cuts and scratches?***

Yes, if they are diluted in water so they are not too strong, but you should not rely on this in the case of a bite wound – you should always have your cat examined by your vet just in case.

Street Accidents

STREET ACCIDENTS ARE THE MAJOR CAUSE OF death in young animals and serious injury in cats of all ages. Cats are explorers by nature. Outdoors in the city or suburbs, they are constantly crossing and recrossing busy streets. There is not much any driver can do to avoid an accident if a cat suddenly darts directly into the car's path, especially if it is dark or raining.

A cat that has been hit by a car will usually run away if it can. It may disappear for hours or even days. If your cat does make it home, it may not at first be obvious that it has been injured in an accident. The cat may merely look dishevelled, wet, or muddy, possibly with oil marks on its fur. On closer examination, however, grazes or areas of skin loss will indicate that it has been in an accident. Another sign is frayed claws (if they were exposed when the cat landed back on the road). The cat is also very likely to be limping or unable to use its hindlegs or tail.

What To Do

If your cat has been hit by a car, your first priority must be to get it to an emergency animal clinic. If this cannot be done immediately, carry out any first-aid measures that may be needed (see pages 134–137), but do not delay getting treatment. Always keep a blanket, the cat's carrier, and a first-aid kit handy, especially if you live near a busy street. This precaution will save precious minutes in an emergency. If a carrier is not available, lift the cat gently onto a firm object such as a tray or large road atlas, trying to avoid unnecessary handling. Wrap the cat in the blanket or a coat or jacket to avoid being scratched or bitten. Even if the cat does not appear to have

▼ *A secure bandage will stop your cat from scratching or licking a wound or pulling out any stitches. For any injury other than a minor one, your cat should be seen by the vet at several stages throughout its recovery.*

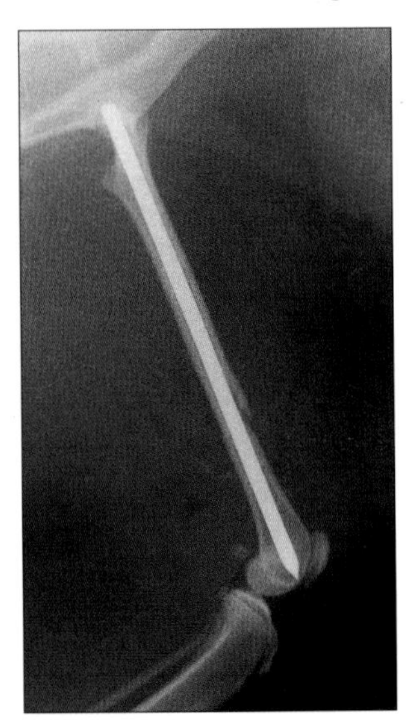

▲ *This X-ray shows a broken leg following a street accident. The vet has inserted a metal pin to immobilise the fractured bone while it heals.*

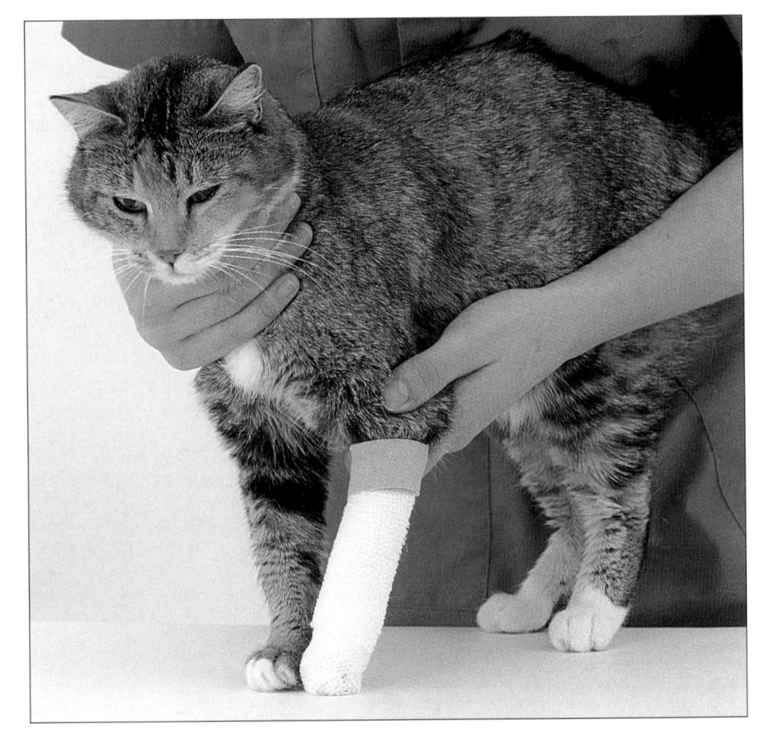

any major injuries, you should still take it to the vet for examination. Internal injuries may not be evident to you, but they can prove fatal if rapid treatment is not given.

If your cat is missing and you fear it may have been injured or killed, search the local area carefully. Injured cats hide themselves away, so look in out-of-the-way places. Circulate a description of your cat to the local animal hospitals, animal shelters, and radio or television stations, and put up notices around your neighbourhood. Finally, an inquiry to the street-cleaning department may, sadly, bring news of a cat's body having been picked up from the gutter.

Common Injuries

Any cat that comes into contact with a moving vehicle or the road is likely to suffer extensive, life-threatening internal injuries. For example, a cat that is having great difficulty breathing may have damaged lungs. If the bladder is ruptured, urine will begin to gather in the abdomen; if the diaphragm has a similar injury, the abdominal organs will move forward into the cat's chest, squashing the lungs. Internal injuries are particularly dangerous because their effects may be delayed by several hours and so overlooked until too late. The cat often suffers delayed shock.

Head injuries must be treated immediately to prevent possible brain damage. Also carrying a grave prognosis are spinal injuries, which often result in weakness or paralysis of the hind legs. This is especially the case when the cat's bladder and bowel function are impaired. If the base of the tail has been damaged, amputation of the whole tail may be the best solution for the cat.

Treatment of a limb fracture may involve an external cast or surgery to immobilise the fracture with a metal pin, plate, wires, or an external scaffoldlike device, depending on whether it is single or multiple, simple or complex. Jaw fractures are common but can usually be repaired. Intensive nursing care and the use of a feeding tube are usually necessary until the cat is able to eat again. 'Degloving' injuries, in which the skin is pulled off a lower limb or tail, can be particularly difficult to deal with. Long periods of repeat dressings followed by reconstructive surery are often required to heal this injury.

Preventing Accidents

Any cat that is allowed outside is at risk of a traffic accident. However, there are a few things you can do to reduce the risk:

- Neuter your cat to reduce wandering.
- Make sure it wears a reflective collar.
- Have it wear an identity tag so you can quickly be contacted if it is injured. An implanted electronic microchip cannot be broken or lost.
- Encourage young cats to play within the garden and do not let them out before meals. The pet door should not face onto the street.
- Build an enclosed run to allow the cat to take exercise without danger.
- Check under your car before you drive away. Cats will sit under cars or even climb in close to the engine for warmth.

● ***My cat Jack is only 2 years old and has already been run over twice. Is it likely to happen again?***

Cats are much more likely to be hit by a car in their first two years of life than later—as they get older, they seem to become more cautious and streetwise. However, a street accident is always a possibility if your cat has freedom to wander outdoors.

● ***My tabby cat Tinker has just been hit by a car. He has a paralysed tail but seems fine otherwise. However, the vet is not sure whether he may be incontinent. Why is this?***

If the tail is caught under the car's wheel and pulled, the nerves at the base are stretched and cease to work. This damage can extend to the nerves that control the bladder and bowels. This may mean that the cat is unable to pass urine, and when the bladder is full, it trickles out continuously. It is difficult to assess bladder function in the early stages after an accident. I'm afraid that euthanasia is often recommended if bladder function fails to return after a week or two.

● ***My vet has advised letting my cat Sinbad rest after his accident to allow his injuries to heal. How can I get him to rest?***

The best way is to place the cat in a small wire-framed crate that allows him room to lie down, stand and turn around without restriction. Line it with a blanket or washable acrylic fleece. If your cat is very active, it may be necessary to give him short-term sedation.

First Aid and Emergencies

WHEN A CAT SUFFERS A SERIOUS INJURY OR THE sudden onset of illness, your first priority is to keep calm (panic will only waste valuable time). Handle an injured or frightened cat carefully to avoid being bitten – cat bites easily turn septic. Attempt first aid only if you are certain of what you are doing; do not allow your attempts to delay you from getting professional help. The basic steps you can take are outlined here, but remember that some first aid – such as artificial respiration and cardiac massage – can be hazardous and should be attempted only when a vet is not within immediate reach:

- Make sure the cat can breathe. Remove any debris from its mouth and gently pull the tongue forwards.
- Place the cat in a recovery position, with its left side uppermost, and keep it warm.
- Check for a pulse (see picture, right). If there is none, check for a heartbeat on the left side of the cat's chest just behind the elbow. Give cardiac massage if necessary.
- Stop any bleeding.
- Put the cat in a secure carrier and transport it to the vet as quickly as possible.

Moving an Injured Cat

Holding the cat's scruff with one hand and supporting the rump with the other (see below), gently lift the cat into a cat carrier lined with soft material. Wrap the cat in a large bath towel, leaving only the head exposed, to avoid being scratched. Lift a seriously injured cat on a rigid object such as a metal tray or board – a road atlas is ideal if you are in the car – and avoid unnecessary movement in case there are broken bones or internal bleeding. This is particularly important if you suspect the cat has spinal injuries. After you have warned the vet of your imminent arrival, drive carefully to prevent jolting, which will cause pain and may do further damage to the animal.

▶ *Correct lifting is crucial to minimise any injury to the cat.*

Artificial Respiration

If a cat inhales blood, vomit, or water following an accident or drowning, it may stop breathing. Send someone for help and attempt artificial respiration:

1. Pull the tongue forwards and clear any obstruction from the mouth. In the case of drowning, gently hold the cat upside down by the ankles of its hind legs and swing it between your legs. This will clear the lungs and stimulate breathing.

2. For all other accidents, lay the cat down so that it is lying on its right side and extend the head and neck forwards to give a free airway.

3. With the cat's mouth closed, cup your hands around the cat's muzzle and breathe into its nostrils for about 3 seconds. Pause for 2 seconds, then repeat. Continue until the cat starts to breathe by itself.

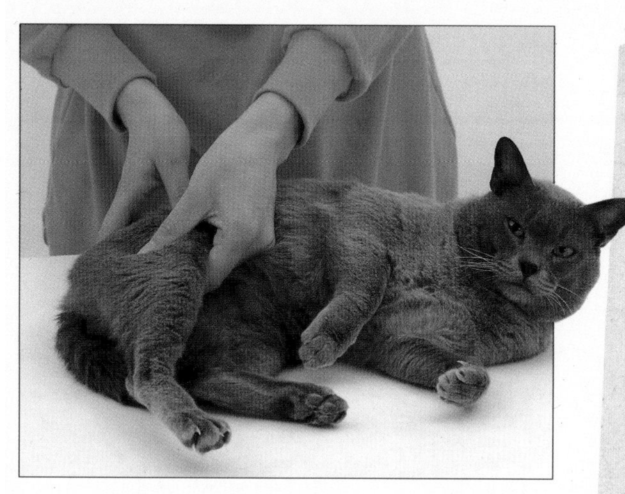

▲ *Check for shock by taking the cat's pulse. You can feel it on the inside of the hindleg. In a healthy cat the rate should be 160–240 beats per minute; it will be higher than this if the animal is in shock.*

The First-Aid Kit

It is sensible to keep a first-aid kit for your cat in the house or car. Many veterinary centres supply special first-aid kits, but you can make your own containing the following items:

- ✓ Bandages of different widths, including a self-adhesive bandage.
- ✓ A gauze pad and sterile dressings.
- ✓ Antiseptic cream.
- ✓ Cotton wool.
- ✓ A length of cord or tape to make an emergency tourniquet.
- ✓ Scissors and tweezers.
- ✓ A small blanket to keep unconscious cat warm and to use for lifting.
- ✓ Disposable gloves.

Cardiac Massage

If the cat's heart has stopped, you should give cardiac massage immediately.

1. With the cat in the recovery position, put the flat of one hand against the cat's left side, just behind the elbow.

2. Spread the hand around the largest part of the chest, with the thumb on one side and the fingers on the other. Pump the rib cage rapidly and firmly three times, then blow into the cat's nostrils: pump, pump, pump, breathe, allowing less than a second for each pumping action.

3. Repeat this sequence about 15–20 times a minute until you feel a heartbeat, then stop pumping. Continue the mouth-to-nose resuscitation alone. Get someone to help you rush the cat to a vet.

Iliac Thrombosis (Saddle Thrombus)

Suspect this if the cat begins to cry continuously, cannot move its hindlegs, has difficulty in breathing, and the gums and tongue are bluish.

1. Lay the cat on its right side, pull the tongue forwards, and extend the head and neck for air.

2. Gently place the cat in a carrier and rush it to the vet.

Electrocution

This can happen to a cat who chews through a live electric cable. This can cause severe electrical burns to the mouth and lips, and in some cases cardiac arrest and death.

1. Switch off the current before attempting to move the cat. Alternatively, use a nonconducting pole (such as a wooden broom) to push the cat away from the electrical supply.

2. If the cat is unconscious, contact a vet at once. Keep the cat warm; check heartbeat and breathing. Use artificial respiration if needed.

3. Rush the cat to the vet.

Bleeding

Unchecked bleeding can quickly lead to shock and death. Rush the cat to the vet as quickly as possible.

Surface wounds

1. Press on the bleeding point with your thumb, or place a wad of cotton wool or gauze against the wound and bandage it tightly. Apply another dressing on top if bleeding continues.

2. For a profusely bleeding leg or tail, use a tourniquet. Tie a narrow piece of cloth (not string or elastic) tightly between the heart and the wound. Use a pencil or small stick inside the knot to twist it tightly enough to stop the bleeding. Tourniquets should not be left on longer than 15 minutes at a time.

Internal bleeding

Suspect internal bleeding if the cat has been in a road accident, had a serious fall, or if the gums become pale and the cat is lethargic.

1. Keep the cat quiet and warm.

2. Minimise movement and seek immediate medical attention.

Cuts and Bites

If a minor wound is visible, you can treat it yourself, but have bites and serious cuts examined by a vet in case of infection.

1. Remove any debris from the wound.

2. Bathe in warm water.

▲ *Minor visible cuts can be treated by bathing in warm water. Clean with a bactericidal soap or diluted hydrogen peroxide. Bites should be treated by a vet.*

Nose Bleeds

May be caused by a blow, violent sneezing, or a medical problem.

1. Keep the cat quiet.

2. Hold ice packs on the nose to stop the bleeding. (A pack of frozen vegetables will do.)

3. Contact your vet if the bleeding persists.

Burns and Scalds

A quick response is needed to limit damage to the skin.

Boiling water or hot oil

1. Cool the skin down by sponging immediately with cold water. Do not apply any ointments or lotions.

2. Apply cold compresses such as a wet cloth or an icepack.

3. Take the cat to the vet immediately.

Chemical burns

Caused by corrosives such as paint stripper.

1. Sponge the area gently with lots of water to remove all traces of the chemical.

2. Consult a vet at once.

Contaminated Coat

Caused if a cat falls or rolls in a spillage such as tar or oil, or after a street accident.

1. Clean the coat with undiluted liquid detergent. Do not use paraffin or similar.

2. Rinse thoroughly with water.

3. If the cat has licked the coat, gently flush the mouth out with cotton wool soaked in lukewarm water. Contact a vet at once.

Objects in the Mouth and Throat

Sharp objects such as fish bones or needles can stick in the mouth or at the back of the throat. The cat salivates and paws its mouth.

1. To avoid being bitten, place a pencil or other wooden implement across the cat's mouth between the upper and lower jaws.

2. Remove the object with your fingers or with a pair of tweezers if you can see it.

3. If you can see but cannot reach a fish bone, or cannot see it, call the vet immediately.

Fish Hooks

These can lodge inside the mouth or in the skin. Do not try to pull one out; it will only tear the cat's flesh. If fishing line leads into the mouth, leave it in place and take the cat to the vet.
1. If the hook is accessible, either push it on through the skin if possible, or cut the hook in two with pliers and then push the barbed end through to the outside.
2. If you can't remove it, take the cat to the vet.

Bee and Wasp Stings

Usually around the mouth or feet. The cat paws or licks at the site of the sting. If the mouth or throat swells, contact a vet at once.

Wasp stings: Wasps withdraw their stinger and fly away. The sting is alkaline, so bathe the area with a diluted acid such as vinegar.

Bee Stings: Bee stingers are left in; remove with tweezers. The sting is acid, so bathe with alkaline such as sodium bicarbonate (baking soda).

Heatstroke

Heatstroke occurs rapidly in cats, often proving fatal in minutes. The first signs are rapid, heavy breathing. The cat becomes distressed, salivates, gasps for breath, and collapses. Never leave a cat shut in a car or other small area such as a greenhouse or conservatory on a sunny day without adequate ventilation.
1. Take the cat out of the heat.
2. Immerse in or douse with cool or tepid water until the panting and gasping stops – this should be within 5 minutes.
3. Offer water to drink.
4. If the cat does not recover rapidly, rush it to the nearest vet.

Accidental Poisoning

This occurs occasionally if a cat eats a toxic substance such as slug bait. Poisonous plants are a potential hazard. Some poisonous substances can be inhaled or absorbed through the skin. The most common signs of poisoning are heavy salivating, sudden violent vomiting, loss of coordination, and convulsions.
1. Contact your emergency centre immediately and tell the vet (if you know) what the cat has ingested or been in contact with. Keep the cat warm and quiet.
2. Do not induce vomiting unless instructed to do so by a vet. Follow the advice given.
3. If an antidote is available, rush the cat to the veterinary emergency centre for treatment.

Common Poisons and Their Effects on Your Cat

SUBSTANCE	SIGNS OF POISONING	ACTION
Alcohol	Depression, vomiting, collapse, dehydration	Call your vet immediately. Identify which kind of alcohol
Antifreeze: Cats like the taste	Loss of coordination, vomiting, convulsions	Call your vet immediately. An injection may be effective
Disinfectants and cleaners: Cat walks in spillage	Violent vomiting, diarrhoea, staggering, twitching	Call your vet immediately. Identify which kind of chemical
Insecticides: Can be inhaled or by contact	Muscle twitches, drooling, convulsions	Call your vet immediately. There is no specific antidote
Painkillers (aspirin, etc.)	Loss of coordination, loss of balance, vomiting, blue gums	Call your vet immediately
Rodent poisons (arsenic, strychnine, warfarin): Cat eats poisoned prey	Restlessness, stomach pain, vomiting, bleeding, diarrhoea	Call your vet immediately. An antidote may be available
Slug and snail bait: Cats like the taste	Profuse salivating, muscle twitches, vomiting, diarrhoea, loss of coordination, convulsions	Call your vet immediately. Prompt treatment may be effective

Cat Breeds

CATS HAVE BEEN DOMESTICATED ONLY RECENTLY, and selective breeding was much slower to take off than for dogs. As a result, cats remain closer to their origins and the range of domestic breeds is relatively small. The average cat weighs between 3.5–6kg (6–12lbs) , though the largest breed, the Ragdoll, averages 10kg (20lbs).

Originally all cats had short hair, but as some animals migrated north of their origins, they developed long coats to shield them against cold winters. Coat length and type are the chief distinctions between cat breeds. Body type is another: the long, slim Siamese or Oriental type and the shorter, muscular type, both associated with shorthaired breeds; and the long, solid type most commonly seen among longhaired cats. Each of these types has a different geographical origin and a typical personality. Cats with the Siamese build but without a Siamese pedigree are often described as 'foreign'. Feline grace and beauty are seen in crossbred cats as well as in pedigree animals.

Pedigree Cats

CATS EVOLVED SLOWLY OVER THE 4000-year history of their domestication, from the original small, spotted African wildcat to the wide range of shorthaired and longhaired cats found across Europe and Asia by the 17th century. This diversification occurred naturally, as cats adapted to a variety of climates and to other environmental conditions, such as the size of their prey. Individual types predominated in their home countries, so it fell to world travellers to notice and admire cats they saw abroad, and import a few specimens on their return to their home country. In this way the Persian (see pages 182–85) and the Turkish Angora (pages 186–87) came to be the forerunners of most longhaired cats in Europe.

▲ *The Siamese was one of the earliest pedigree breeds and has remained hugely popular. Fans of the breed love it equally for its elegant appearance and its outgoing personality, though some Siamese can be highly strung.*

Selective breeding and the introduction of registered pedigrees did not begin seriously until the late 19th century, when the first cat competitions were held – first in England, then the United States and Europe. Among the first recognised breeds were Siamese, Persians, Manx cats, Abyssinians, Angoras, and Maine Coons (in the United States only). Today there are more than 50 pedigree groups, though some of the more recent are controversial and not universally accepted. There are also some international differences: for example, the Oriental Shorthair (see pages 164–65) is considered a separate breed from the Siamese in the United States but not in Great Britain.

Why Get a Pedigree Cat?

Many people are attracted to a particular breed for its distinctive appearance. Others choose breeds for their different personalities, which have also developed through selective breeding. In general, slim Siamese-type cats tend to be noisy extroverts, while large, solidly built cats, such as Persians and Maine Coons, are quieter and less outgoing. However, although fans of a particular breed may argue there are endless subtle differences that set it apart from all others, it is questionable whether these are traits of the whole breed or of individual cats.

You may be thinking of showing your pedigree cat, but this should not be undertaken without careful consideration. Purchasing and caring for a show-quality cat is a considerable investment of time and money, and there is no guarantee that a young kitten will turn into a champion adult. It takes a particular temperament to allow a cat to endure long trips to unfamiliar places and the indignity of being prodded and lifted by complete strangers. You should compete only if both you and your cat genuinely enjoy it, and if you can take the inevitable losses in your stride.

● *I want a pedigree Birman, but they all seem to be out of my price range. Any suggestions?*

Ask your vet or someone in the pedigree cat world to help you find a Birman breeder. Ask for a healthy cat but not one of show quality. You may be able to get a cat that is unsuitable for showing because it doesn't match the breed standard. The cat will probably be sold without a pedigree certificate, and you may have to agree in writing to neuter it, but it should be affordable. Alternatively, the breeder may offer you terms under which you allow your cat to be mated and the pick of the kittens to be returned to the breeder for sale. Anything you sign in either case constitutes a legal contract, so be prepared.

● *Is inbreeding among cats as unhealthy as it is in humans? Does it matter if I'm looking at a pedigree cat whose mother was mated to her own brother?*

In general, inbreeding is not healthy, but it is the only way to concentrate desirable inherited traits. However, this may also concentrate undesirable and even lethal traits, so you should question the breeder about why he bred this pair. If you are looking at a new breed, inbreeding is understandable, but you should find out if the kittens are likely to have any health problems. Contact the relevant breed association for advice.

If You Want A Pedigree Cat

- ✓ Ask your vet, friends, or a pedigree association to recommend a breeder.
- ✓ Always visit the breeder in person to choose your kitten or cat.
- ✓ Visit more than one breeder and compare their facilities and prices.
- ✓ If you intend to show the cat, take along someone who knows the breed and can advise you about your selection.
- ✓ Make sure that both parents are registered with the relevant breed registry.
- ✓ Ask about any hereditary diseases in either one of the parents' lines (see Breed Profiles, pages 146–199).
- ✓ Agree in advance what you will do if your new kitten is sick or fails to settle in.
- ✓ Your kitten should have been registered already (at 5 weeks); be sure to complete and return the transfer documents.

▼ *A Persian queen with young kittens. Persians have a long history of showing and are still the most popular of the pedigree breeds.*

Nonpedigree Cats

THERE ARE MORE THAN 100 MILLION PET CATS worldwide. Of these, the vast majority are nonpedigree or crossbred cats: their parentage may be unknown, or if it is known, neither of the parents is a registered pedigree. Up until the late 19th century, people kept cats in order to keep their houses and barns rodent-free; looks were not a consideration. However, as every proud owner knows, a healthy, happy crossbreed in the prime of its life can be every bit as magnificent in appearance and manner as a pedigree cat.

• *Everyone comments on how beautiful my 2-year-old calico (tortoiseshell) cat is. Is there any way I can show her?*

Some cat show organizations have a special category that is open to nonpedigrees. Contact your local cat association to find out which ones to apply to.

• *Our white cat, Snowy, gave birth to five kittens fathered by the white cat next door. They are all different colors—one white, two tabby, one tabby and white, and a black. How did this happen?*

The gene for a white coat is dominant over other color genes. Two white cats may each carry recessive color genes, and kittens with two recessive color genes will appear as different colors.

• *Can wild cats breed with domestic cats?*

Not all of them. The African wild cat cannot interbreed with domestic cats, but the European wild cat can. Indeed, interbreeding with feral cats (domestic cats that have escaped or been abandoned to live in the wild) means that there are probably very few colonies of true wild cats left in Europe.

Moderate Build, Many Colors

Because the gene determining short hair is dominant, most crossbred cats are shorthaired, but there is no "standard" crossbred type. They come in every imaginable variety of coat and color. Many crossbred cats are tabbies, which is the variety closest to the cat's ancestors among African wild cats. The mackerel or striped tabby pattern is the original, but the classic (blotchy) tabby pattern is more common. Most rare is the spotted tabby, now being selectively bred in new pedigree lines (such as the Ocicat; see pages 176–177) that aim to achieve a wild look. Solid colors also abound: black, white, marmalade (ginger), and blue. The marmalade coloration is sex-linked, being carried on the X chromosome, and marmalade males outnumber females by about 2 to 1. Conversely, the calico or tortoiseshell pattern of orange and black is only possible in females (with very rare exceptions in sterile males). White is common in crossbreeds, both on its own and in combination with other solid and tabby colors. Siamese-style points are rarely seen in crossbred cats but can certainly occur.

Apart from color and coat, crossbred cats differ much less from each other than pure breeds do. Most have the moderate build that is typical of American and British Shorthairs (see pages 146–149), being neither slender like the Siamese nor large and heavy like the Persian or Maine Coon. Although random-bred cats from tropical climates tend to have a somewhat sleeker form than others, and those from cold climates are comparatively stockier, they have not acquired

Advantages of a Crossbreed

- ✓ Whether you are looking for a kitten or an adult, there is likely to be a huge selection to choose from in your area.
- ✓ You will not need to contact a breeder. They can be obtained from shelters or private homes.
- ✓ You have to pay very little. Most crossbreeds are given away.
- ✓ Crossbred cats are robust and long-lived.
- ✓ They suffer from fewer inherited health problems than pedigree cats.

▲ *Nonpedigrees are generally stocky and are found in a wide range of colours. Shown here are tabby (left and front), black-and-white (centre) and ginger (right).*

acquired the extreme lines that have been introduced into pedigree lines by selective breeding. Wedge-shaped heads and flattened faces are unusual in a crossbreed but can appear if the recent family tree of one of the parents includes a cat with Siamese or Persian genes.

The World's Favourite Family Pet

Random breeding means that the nonpedigree cat does not have a definitive appearance or temperament. Yet the character traits of the domestic shorthaired cat make it universally loved and admired. Cats are wonderful companions and enjoy being part of a family, but still retain much more independence than domestic dogs. They can adapt to an indoor existence but will make the most of any freedom offered (deliberately or otherwise), for it must be said that the domestic cat is a relentless predator of small rodents and birds. Even when well fed, most crossbred cats will persist in bringing hunting trophies home, faithful to their thousands of years of heritage as pest-control specialists.

The crossbred cat has hybrid vigour – nature's way of selecting the fittest and most successful animals. Crossbreeds have much lower concentrations of undesirable genes (one of the problems of selective breeding; see pages 140–141). With proper care it is robust and should live a long life. The typical crossbred, if you choose carefully, is a beautiful, intelligent, playful, low-maintenance companion with an independent streak. It will be a devoted and loving member of your household. Who could ask for more?

Shorthaired Cats

Lilac Burmese

Because the gene for short hair is dominant over the gene for long hair, shorthaired cats remain the great majority among domestic cats. They reigned almost supreme until the end of the 19th century, when selective breeding began to produce a wider range of longhaired cats than existed naturally.

A cat's coat may consist of three types of hair: the longer guard hairs of the top coat and two types of downy undercoat hairs, thicker awn hairs and wrinkly, softer true down hairs. A typical shorthaired cat will have guard hairs some 4.5cm (1¾in) long, and the thickness of the coat varies according to the mix of the three types. British, European and American Shorthairs have a very dense water-resistant single coat; 'foreign' types like the Russian Blue have a plush double coat with a very dense, soft undercoat; while the Oriental breeds have a very short, silky, glossy, close-lying single coat.

A short coat is easier to groom than a long one, both for the cat itself and for the owner. It rarely takes more than a few minutes a day, and stroking by hand can be helpful if the brush or comb is temporarily out of reach. Tangles are minimal, shedding is not a big problem, and parasites such as fleas are much easier to spot because there is less fur to hide in.

Whether your taste runs to sturdy no-fuss farm cats, the ultra-refined Siamese and its crosses, or to new 'wild' looking types such as the Ocicat, you will almost certainly find a shorthaired cat that is right for you.

Scottish Fold (right)
Seal Point Siamese (below)

American and Exotic Shorthairs

Tail *is thicker and slightly longer than in the British and European breeds, with a rounded tip*

Head *is oval, where the British Shorthair's is round*

Body *is big and powerful but athletic rather than massive*

Legs *are medium in length but powerful*

▲ *The American Shorthair is an undemanding family pet but will be much happier if allowed outdoors. The coat can be very striking, as in this blue tabby.*

THE OLDEST RECOGNISED AMERICAN BREED, the American Shorthair, is very similar to British and European Shorthairs and just as popular in its native land. According to some sources, the first domestic cats arrived in North America by ship with European settlers in the early 17th century, and American Shorthairs developed from these. Non-pedigree shorthaired cats had certainly been available to keep down the rodent population of North America long before a pedigree was introduced, but when cat shows began to proliferate during the 19th century, American judges and breeders favoured the more flamboyant local longhairs, the Maine Coons. It was not until 1965 that an American Shorthair won the top championship in its own country.

Breed Profile

Life expectancy:	14 years
Adult weight:	3.5–8kg (8–18lbs)
Average litter size:	4

Temperament: Relaxed and friendly with people. Prefer access to outdoors. Proficient hunters. Fur is easy to groom but needs daily combing.

Colours: Self and tortie colours; smoke, shaded, tipped, tabby, shaded tabby and bicolour; smoke, shaded and tipped bicolour; tabby bicolour; silver tabby bicolour.

KNOWN HEALTH PROBLEMS

There are no special health problems associated with the American Shorthair.

From Common to Pedigree

The pedigree American Shorthair line began at the turn of the 20th century with an imported British Shorthair. Additional imported cats were mated with the local shorthaired cats to develop the new breed. Initially called Shorthairs and subsequently Domestic Shorthairs, their name was changed to American Shorthair after a silver tabby shorthair won Cat of the Year in 1965. Originally, any shorthaired cat that conformed to the breed standard could be registered as a pedigree, contributing to a large gene pool, but this is no longer allowed by any cat registry.

● I want to get an American Shorthair, but there are poisonous snakes where I live. Will the cat try to hunt snakes ?

Your American Shorthair is likely to develop into an efficient hunter – but will usually turn its attention to prey species only: rodents such as mice, shrews and rats, rabbits and small birds. It is very uncommon for a cat to attack snakes, and snakebites are much rarer than they are in dogs.

● Nero, our black Shorthair, has begun scratching all the time. I can't see any fleas on him; what else is it likely to be?

It is almost impossible to see either fleas or flea dirt (minute black specks) on a black cat. Dab a wet white cloth or tissue over Nero for a second or two. If it picks up tiny black dots that smear red when damp, this is flea dirt. If none are present, take him to the vet; the problem is likely to be something other than fleas (for more on skin problems, see pages 92–95).

● I have just acquired an Exotic Shorthair kitten, Geraldine. Will she need grooming as often as a longhaired cat?

The simple answer is yes, but the grooming won't take nearly as long. A daily combing or light brushing will be quite sufficient, and you are unlikely to have any matting or difficult tangles to deal with.

▲ *An Exotic Shorthair has a fluffy coat and a Persian face. It combines the best personality traits of Persians and Shorthairs, and makes an ideal family pet.*

The American Shorthair has developed along slightly different lines from its British and European cousins. Its build is less stocky, with a more oblong head, longer nose, larger ears and longer legs. Its body is both leaner and more powerful, suggesting that its ancestors evolved to match the pioneer lifestyle of early generations of their owners. Pedigreed American Shorthairs are bred to consolidate these qualities without exaggerating them – a mistake typically made with breeds that became popular earlier, when the science of genetics was less well understood.

The American Shorthair's coat is thick, hard and dense, making this cat as well equipped as the Maine Coon to withstand hard winters. The coat comes in more than 30 colours and patterns, but the best known is the tabby – either the classic brown tabby or the strikingly marked silver tabby. Calico (tortoiseshell and white) and bicolours are also popular, and new colours are still being developed by breeders.

The American Shorthair is a ruthless hunter, but its sturdiness and laid-back attitude make it an ideal pet for a lively household with children. Although it is not usually demanding or destructive, it will not be at its best if confined indoors.

Exotic Shorthairs

This breed was developed in the mid-1960s by crossing Persians (see pages 182–185) with American and other shorthaired cats. The product is a shorthaired cat with a softer, fluffier coat and a rounder, shorter head with a Persian face. As a result, Exotic Shorthairs can have some of the defects of Persians, such as a tendency to blocked tear ducts and dental problems. However, they also share the Persian's gentle nature.

The coat occurs in all colours found in American Shorthairs and Persians. Pedigree cats must have only Exotic Shorthair, Persian, or American Shorthair lineage. The show standard is identical to the Persian's except for the coat.

British and European Shorthairs

BRITISH AND EUROPEAN SHORTHAIR CATS MAY have originated from felines brought into the British Isles by the Romans. They almost appear to be a cross between a cat and a teddy bear, rounded and cuddly, but they are also one of the best hunting breeds around. They are stockier, heavier and much quieter than their Oriental counterparts. Although their dense fur, stocky appearance and lack of exotic origin have limited their popularity in cat shows, their genes are found in abundance among the huge number of nonpedigree shorthairs kept as pets worldwide.

British Shorthairs won early shows in Britain in the late 1890s before being eclipsed by the rise in popularity of longhaired Maine Coons and Persians. All Shorthairs (see pages 146–147 for the American Shorthair) have the same origins and initially had the same breed standards, but in 1982 the European Shorthair was given separate status. However, no doubt because of its close resemblance to the British and American Shorthairs, it is still rare worldwide.

Nearly Identical Cousins

In spite of their separate breed standards, British and European Shorthairs are virtually identical. Both are well-balanced, compact cats showing a good depth of body and a full broad chest. They have short, strong legs, round paws and a thick tail with a rounded tip. The head is large, round and wide in span between small ears. Their eyes are large and round, with no hint of an Oriental slant, and usually copper, orange, or deep gold in colour. The face and nose are short, broad and straight with a shallow nose break. Both breeds have a firm chin with a level bite.

▶ *British and European Shorthairs are almost identical and come in many colours; the blue-cream and the bicolour black and white are shown. The two groups almost disappeared in the first half of the 20th century but have been revived since the 1950s. Most nonpedigree shorthairs are part British Shorthair.*

Breed Profile

Life expectancy:	14–15 years
Adult weight:	3.5–7kg (8–16lbs)
Average litter size:	4

Temperament: Quiet, self-sufficient, not usually temperamental, though they may dislike being picked up. Fairly easy to own. They enjoy the company of people but are not demanding – a good breed for beginners. Suitable for living in a flat but enjoy safe access to the outside. They are efficient hunters. The coat is easy to groom and should be combed daily.

Colours: All solid colours, tortie, tabby, smoke, tipped and bicolour.

Known Health Problems

With the exception of deafness, an inherited trait often present in white cats, there are no particular health problems associated with either the British or the European Shorthair.

Coat is thick to protect against cold, rain and skin injuries

▶ *The tabby pattern is popular in both pedigree and nonpedigree shorthairs, and can be distinguished by the M-shaped 'frown marks' on the forehead. This British Shorthair's silver coat contrasts with its deep-gold eyes. The coat needs very little maintenance.*

The coat of the Shorthair is short, dense and crisp. Having evolved in a typically cool, wet climate, the coat is weatherproof, and a little rain or snow will not deter these cats from going outdoors in pursuit of their favourite activity, hunting. Daily grooming with a comb and lots of stroking will keep the coat in prime condition.

Cats of Many Colours

Shorthairs are available in almost all colours. The classic solid colours are blue (still sometimes given separate breed status as the British Blue), black, red, cream and white. Chocolate and lilac have recently been added to this list, and the classic tabby pattern is still very popular – but there can be very few cats more striking than the silver tabby, with its dense black markings on a silvery background. Tortoiseshells and the bicolour tortie and white are common only on females as the tortoiseshell pattern is linked to gender. Tipped, smoke and bicolours are all available, and the Spotted Tabby pattern is now appearing in a wide variety of its own colours. Most Shorthairs have gold or copper eyes, but white cats may have blue or odd-coloured eyes (one blue and one of another colour).

Closely related to the British Blue is the Chartreux, a French cat said to have originated in a monastery in the 16th century. The differences

▲ *The Chartreux, a French shorthaired cat, looks identical to a British Blue Shorthair, but it was originally larger in build. Its lineage goes back nearly 500 years.*

now seem to be merely a less round head and a slightly more silvery coat.

Both breeds have imported the Siamese pattern, in which colour is restricted to points on a pale coat. The background colour is darker than in the Siamese and tends to deepen with age. The variation is known as a Seal colourpointed British Shorthair in the UK and as a Seal Point Shorthair in Europe.

Don't Pick Them Up

As a breed, shorthairs are robust, affectionate and intelligent. They are not as temperamental as the Orientals (see pages 164–165), nor are they as noisy and demanding. Show handlers and breeders report that they typically dislike being picked up, but that does not stop them from climbing onto your lap for a cuddle – as long as it is on their terms. They are easy to breed and their kittens are equally robust. Unlike most shorthairs, about half of the British group has type B blood, a curious anomaly.

● ***We are expecting our first baby soon and are worried about how our cat will behave. Six-O is a 2-year-old Silver Tabby, and although he is used to children visiting, he has had no contact with babies.***

Most Shorthairs are fairly laid-back, and he should accept the baby without great difficulty. However, you should not let him have access to the crib and do not leave him alone in a room with the baby. Establish a routine that you will be able to continue after the baby is born. For instance, if Six-O sleeps on the bed, change that now. Also, it's very important not to give him less attention once the baby is born.

● ***Camilla, our white European Shorthair, likes nothing better than to bask in the sun. I have been told that she is at risk of sunburn. Is this a joke?***

Absolutely not. White or white-faced cats and those with white ears are very likely to suffer from skin problems caused by too much exposure to the sun. The most sensitive and susceptible parts are the ear flaps and the nose. If they are overexposed, small red crusty areas may develop, and these can lead to cancer. For her own sake, you should try to keep Camilla indoors on sunny days. Ask your vet's advice about using sunblock.

Russian Shorthair

BOTH DIGNIFIED AND BEAUTIFUL, THE RUSSIAN shorthair is considered by many cat fanciers to be the perfect indoor cat. The breed originated near the White Sea in northern Russia and was introduced to Britain during the 19th century. Although it was known in North America in 1900, it did not become established there until the 1950s. Blue was the original coat colour and is emphatically preferred by traditionalists, but black and white coats are also available, especially in Europe and New Zealand.

The Russian is basically a 'foreign' or Oriental type of cat. Its build is lithe, with a finer bone structure and a longer, wedge-shaped head than other shorthaired cats. Despite its popularity, the Russian Shorthair nearly became extinct during World War II, but it was revived using Finnish Blue cats crossed with Blue Point Siamese. This technique was taken up by British breeders and resulted in an extreme foreign look. The show standard was rewritten by 1965 to encourage a return to the original type of Russian Blue.

Breed Profile

Life expectancy:	14 years
Adult weight:	3.5–5kg (7–12lbs)
Average litter size:	4

Temperament: Quiet, affectionate, but shy with strangers. Does not cope well with change. One of the least destructive cats; perfect for living in a flat. Requires only minimal grooming: hand stroking and rubbing with a chamois cloth are sufficient. Some breeders say the coat looks best if it is never brushed or combed at all.

Colours: Blue, black, white.

KNOWN HEALTH PROBLEMS

There are no particular health problems associated with Russian Shorthairs.

▼ *The traditional Russian Shorthair is a blue cat with a silvery sheen to its coat. It makes an elegant companion but is not well suited to active family life.*

Body is long and slender but muscular, with a plush coat

Tail is medium length, tapering to the tip and not as thick as the coat

Head is short and wedge-shaped, with long whiskers

Hindlegs are slightly longer than the forelegs

Paws are small and round, at the end of long legs

Korat

▲ *The silver-blue Korat is a rare cat even in Thailand, its country of origin, and is supposed to bring prosperity and good fortune. It is valued as well for its personality and affectionate disposition.*

AN ORIENTAL CAT HAILING FROM THAILAND, the Korat is considered a good-luck charm in its native land. Its Thai name, 'Si-Sawat', can have the meaning of 'good fortune'; it gets its international name from the region of Korat, a high plateau in the remote northeast of the country, where it has been known since at least the 14th century and is mentioned in the *Cat Book Poems* of the Ayuttha Kingdom (1350–1757).

Although an ancient breed with a long lineage, the Korat has been known outside Thailand and Asia for only a short time. It is thought that the first exported Korat, described as a solid-blue Siamese, may have been seen in Britain in the late 19th century. However, the modern history of the breed began in 1959, when the first pair was imported into the United States from Bangkok. The Korat attracted considerable attention and gained official recognition in 1966. It was not introduced into Europe until 1972.

A Heart-Shaped Head

Although the Korat resembles the Russian Blue in many respects, there are important differences. Like the Russian, it is a strikingly handsome cat with piercing green eyes and a short, silvery-blue coat. The head, however, is not at all like the Russian's. It is much softer in appearance and is heart-shaped when viewed from the front. In profile, it is shorter than the Russian's, and there is a slight stop between the forehead and the nose. The eyes appear a little oversized for the face. Round and luminous when wide open, they appear Oriental when partially closed. The brilliant green eye colour does not appear until the cat is mature, usually at 2–4 years. In kittens

● *Leila, our Korat cat, hates being combed – we think her skin is very sensitive. What can we do to keep her fur in tip-top condition?*

Try a wet chamois cloth. Just polish her coat with it; she will almost certainly enjoy it if you are gentle, and her coat will gleam afterwards.

● *We've just been to see a litter of 11-week-old Korat kittens as we are hoping to buy one. Though the mother was a beautiful cat, her kittens were quite ugly and we were put off from choosing one. Is this common?*

Oddly, Korats seem to go through an 'ugly' phase as they grow up – a bit like the phase a cygnet goes through before it develops into an adult swan. Don't be put off. Your Korat kitten will turn out to be just as beautiful as its mother.

● *I can hardly move around the house without Kee, my Korat, following me everywhere. She is deeply attached, and though I love her attentions, I'm worried that she is bored and lonely when I'm out at work. Would a companion cat help?*

Yes, I think it would be a good idea for Kee. Korats bond very strongly to their human companions and she almost certainly misses you. Two Korats usually get along very well and will entertain each other. Kee will probably become less dependent on you, but her quality and enjoyment of life will improve.

▲ *The rounded lines of the heart-shaped head and the large, luminous green eyes give the Korat a soft and gentle appearance. It confronts the world with an alert and intelligent expression.*

and adolescents the eyes are yellow or amber to amber-green. The large ears are set high on the head, giving the cat an alert expression. Korats' bodies are well-muscled and strong.

The coat is short and lacks an undercoat. It is blue with the guard hairs tipped in silver, giving it an intense sheen. Unlike the Russian Blue's double coat, which is similar in colour but springs up quickly when flattened, the Korat's fine coat lies close to the body and needs only a cloth or comb run over it to keep it smooth.

Korats are quiet-voiced, sensitive, and intelligent cats. Given affection and plenty of attention, they will bond closely with their owners, but they are also demanding and strong-willed and like to have their own way. This can make them stubborn and territorial if they are left to amuse themselves all day. They remain playful until old age and are highly trainable. Though they make good mothers, the litter size is small.

Breed Profile

Life expectancy: 16 years

Adult weight: 2.5–4.5kg (6–10lbs)

Average litter size: 1–3

Temperament: Quiet in voice but strong in personality – demands attention and likes to have its own way. Can be pushy. Gregarious by nature and bonds strongly with humans. Playful and highly trainable.

Colours: Blue is the only colour.

Known Health Problems

GM1and GM2 are rare neuromuscular diseases that occasionally affect Korats. Blood tests will show if they are present.

Devon and Cornish Rex

UNIQUE AMONG CATS, BOTH REX BREEDS HAVE short, velvety, regularly rippled or wavy coats. This, and the close proximity of their origins on the southwest coast of England, suggests that they are variations on one breed, but in fact they are totally distinct breeds founded by two unrelated cats that underwent the same genetic mutation independently, 10 years apart. The same mutation is known to occur in mice, rats, rabbits and horses. The cats' sheer curiosity factor among breeders guaranteed that both lines were continued. The American Cat Fanciers' Association, which recognised the Rex in 1963, only three years after the Devon Rex appeared spontaneously, did not distinguish between the two breeds until 1979, however. A third Rex cat, the German Rex, which is genetically identical to the Cornish Rex, is also recognised.

Both breeds of Rex are slender, with curving backs, disproportionately small heads, and large ears, which sometimes have the effect of giving them an 'extraterrestrial' appearance. But their popularity owes as much to their personalities as to their looks – they are affectionate, intelligent and outgoing. Most cats only wag their tails when angry or worried, but Rexes are reported to wag theirs when they are pleased as well. They are very lively and agile, are generally easy to breed and are usually found to be good mothers.

● ***How much grooming does a Rex need?***

All you need usually do to keep the Rex's coat soft and shiny is give a daily brushing – stroking the cat firmly from neck to tail – or a light rubdown with a cloth.

● ***We would love to have a cat, but our daughter is asthmatic and we are concerned that she may have an allergic reaction to cat hair. Would a Rex be a good choice of pet, given her condition?***

Because Rexes are nonshedding, they make good pets for people with an allergy to inhaled cat hair. However, asthma can sometimes be brought on simply by stroking an animal or contact with its saliva. You should have your doctor run sensitivity tests on your daughter before committing yourself to any cat.

● ***We are moving some way north with Ringo, our Cornish Rex. I know that Rexes feel the cold, so how can we help him cope?***

You will probably find Ringo prefers to stay indoors until he gets used to the colder temperature. A baby T-shirt or small, tight sweater should keep him warm when he ventures out. He is likely to start eating more and put on weight as his own way of coping.

▶ *Small bundles of curiosity. Devon Rex kittens develop very rapidly but retain their mischievous, inquisitive personalities throughout their lives.*

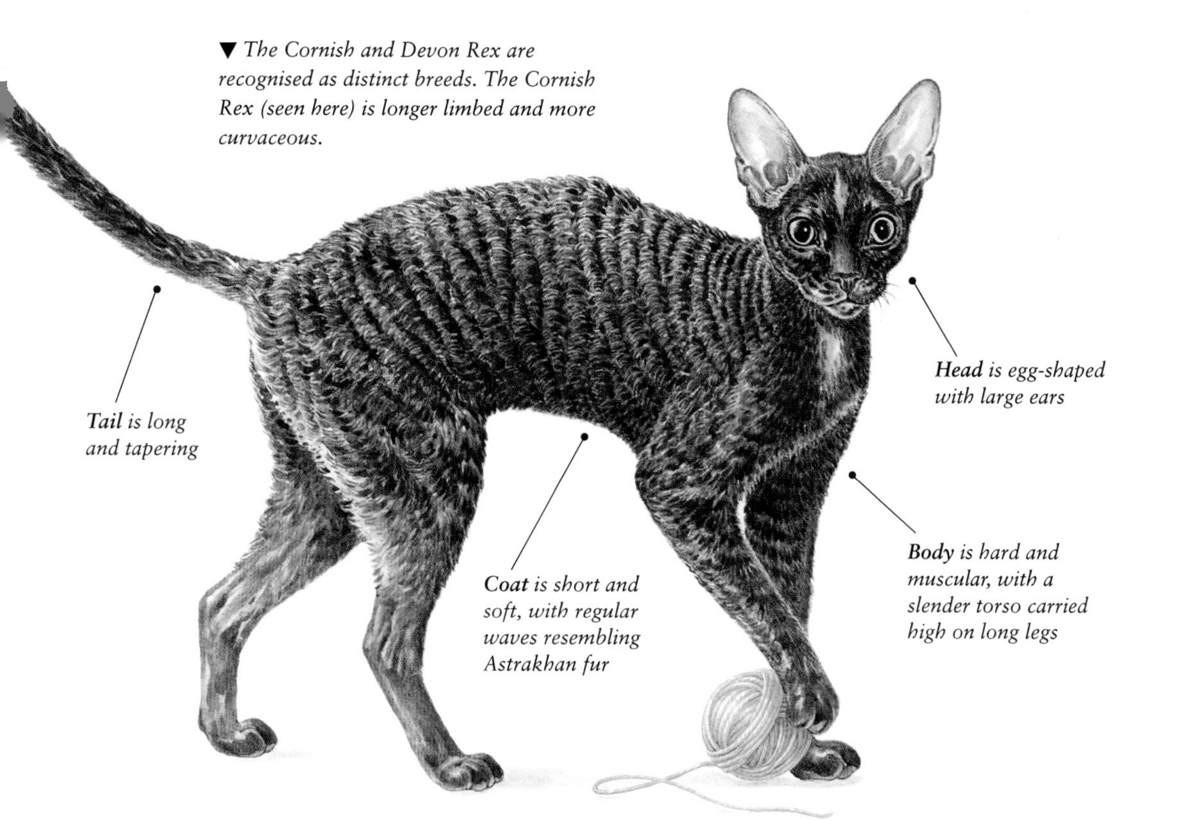

▼ *The Cornish and Devon Rex are recognised as distinct breeds. The Cornish Rex (seen here) is longer limbed and more curvaceous.*

The Poodle Cat

This nickname – which comes from the wavy coat and the tail-wagging habit – is usually used to describe the Devon Rex, which has a slightly curlier, coarser coat than the soft, silky one of the Cornish. An even more significant difference between the two breeds is in the body shape. The head and body of the Cornish Rex resemble a Siamese, with a longish triangular head and oriental eyes. (Siamese were often used as outcrosses when the first imported Cornish Rexes were bred in the US in the 1950s and 1960s.) The body is carried high on the legs, and the arched back is very pronounced. The Devon Rex has a wider, elfin face, with huge, low-set ears, larger, rounder eyes, prominent whiskers and a short nose. Both have long fine tails. Widespread outcrossing means that almost every colour and pattern of coat is found.

Both breeds require minimal grooming and hardly moult at all. Their lean appearance disappears when they move to cold regions as they will put on an extra layer of fat to compensate.

Breed Profile

Life expectancy:	12–13 years
Adult weight:	2.5–4.5kg (6–10lb)
Average litter size:	3–6

Temperament: Playful and affectionate. Good family cat – enjoys company. Easy to breed.

Colours: All colours and patterns, including pointed, sepia and mink.

Known Health Problems

Health problems are rare, but there are some characteristic defects.

Spasticity may appear in Devon Rex kittens at about 3 months. They are nervous, panic easily, have difficulty in getting to their feet, walk with a high-stepping movement and tire easily. The head seems tucked under the chest, so the kittens cannot eat easily and often cough or vomit. The condition is rare and thought to be hereditary.

Gingivitis (inflamed gums) and periodontal disease are quite common.

Hairlessness and a lack of whiskers sometimes occur.

Scottish Fold

THE SCOTTISH FOLD IS AN EXAMPLE OF A spontaneous mutation. In 1961 a shepherd in Scotland noticed that one of the local farm cats, Susie, had peculiar ears that were folded forwards. When Susie produced some kittens with folded ears, the shepherd and his wife acquired one and began to experiment with breeding, crossing it with a British Shorthair. Eventually, at a cat show, they met a geneticist who took one of the current generation of fold-eared kittens in order to investigate the mutation. Her work revealed that the folded ear is due to a single dominant gene, which means that all Scottish Folds must have one parent with the gene. Thus all true Scottish Folds are Susie's descendants.

▼ *The Scottish Fold is essentially a standard shorthair with a medium body, rounded head and distinctive folded ears. It is available in all shorthair colours. Tabbies such as this one have always been popular.*

Relatively early in the breed's history, several Scottish Folds were sent to the United States for study by researchers. The cats quickly caught the attention of breeders there, and the Scottish Fold was accepted for registration in the United States in 1973, attaining full championship status in 1978. Today it is one of the ten most popular cat breeds in the United States. In Britain, though, the Scottish Fold is not accepted for pedigree registration because of concern that the ear folding gives rise to deafness and other problems.

An Owlish Shorthair

The Scottish Fold has a characteristic round face with wide round eyes and the ears folding tightly forward over the head so that the rounded skull can be clearly seen. The overall impression made by the breed is of a sad, owlish look on a cat that otherwise resembles the British or American Shorthair. The Fold's eyes are separated by

Breed Profile

Life expectancy:	14–15 years
Adult weight:	3–6kg (7–13lbs)
Average litter size:	3–4

Temperament: Sweet-tempered and undemanding but needs company. Makes a good family pet. Likes children. Vocalises in a soft voice. Easy to groom. Does well in cold climates and enjoys hunting.

Colours: All colours found in British and American Shorthairs, except chocolate and lilac; all patterns except Siamese and Himalayan.

Known Health Problems

The Scottish Fold, with its ancestry of farm cats, is generally resistant to disease, but the dominant gene responsible for producing the folded ear can cause inherited problems in some cases. Cats showing these faults should not be used for breeding:

Abnormal Ear Folding The ideal of small, tightly folded ears does not always occur in Scottish Folds. Some cats' ears may be so tightly folded that there is hardly any entrance into the ear canal. Other cats may have large, floppy ears. Both these types tend to have poor ventilation of the ears.

Thick Limb Bones Bones in the tail and leg of the Scottish Fold may be too heavy and thick. While not life-threatening, this condition is uncomfortable and leads to disqualification in American shows.

Deafness This was originally thought to be common in Scottish Folds but is now known to be due to the colour gene. It is seen only in white Folds.

● ***We want to get a Scottish Fold cat but have been told they suffer from ear mites. Is this true?***

The Scottish Fold is prone to a variety of ear problems, including mites, due to the structure of the ear. Because the ear folds over forward, it tends to close the entrance to the ear canal somewhat, and this can lead to the ear not being well ventilated. A moist ear is more likely to attract any infection, so you must be attentive to ear hygiene in Folds. If this is likely to cause a problem for you, you should choose a short- and wide-eared breed like the Shorthair, which is less likely to have problems. On the other hand, if you are prepared to keep the cat's ears clean, your Scottish Fold may not have any more ear problems than another breed.

● ***Our Scottish Fold, Ailsa, is a year old now, and we would like to breed her. I've heard that I should not mate her to another Scottish Fold. What are the alternatives?***

If you mate her with another Scottish Fold with folded ears, there is a greater chance of producing tail and limb abnormalities, so you can do one of two things. Either mate her to a British or American Shorthair, or to a straight-eared cat born to a female with folded ears. As the fold gene is dominant, half the litter on average should have folded ears. The others will make good pets or can be used in future breeding programmes. It would be a good idea to join your local cat society to get in touch with other Scottish Fold breeders.

a broad nose, which is short with a gentle curve and sometimes a brief stop as well. The folded ears should be small with rounded tips and set on the head like a cap. The tail should be medium to long, flexible and tapering.

The Fold's coat is short, soft and dense, and is found in most of the patterns and colours recognised in British and American Shorthairs. Longhaired Scottish Fold kittens have been present since the first litters from Susie, but as the longhair gene is recessive, this variety is less common. The only difference is the coat length.

At birth all kittens' ears look the same, but by four weeks of age or so, when all other kittens' ears begin to straighten and stand up, the ears of the folded kittens can be seen to be folding forward and down. There may be a wide range of types of fold; the first members of the breed all had single folds, whereas today's show cats have tight triple folds. Scottish Folds continue to be crossed with British and American Shorthairs to prevent malformations of the skeleton.

Scottish Folds are gentle, undemanding cats. They love family life, but with safe access to the outside they are efficient hunters. They can cope with cold, harsh weather and have the farm cat's resistance to disease. Folds like company and should not be left alone for long periods. Many have the endearing habit of sitting upright and tapping their owner with a paw. They often sleep on their backs with their legs in the air.

American Curl

THE AMERICAN CURL IS A NEW BREED OF CAT, first noticed and developed in Southern California in 1981. It is not a fabricated breed but a result of a breeding programme using cats with a naturally-occurring genetic mutation. This was first observed in a domestic cat, a black longhaired stray called Shulamith whose ears curled back in a very distinctive way. When she was bred by the couple who had adopted her, half her kittens also had curled ears. Thus the new breed was born. Interestingly, these first kittens included curled colourpointed kittens.

Because Shulamith, the first of the line, was a longhaired cat and the longhair gene is recessive, all early Curls were longhairs. A shorthair version was soon developed, but matings often produced longhaired kittens due to the concealed longhair gene. In the show standard in the US, longhairs and shorthairs are judged to an identical standard except for the coat.

The distinctive ears of the American Curl are its most noticeable feature. They turn backwards in a crescent shape due to a genetic change in the cartilage of the ear. The gene that controls this characteristic is an incomplete dominant gene, which means that a cat that has one gene for straight ears and one for curled ears will have curled ears. The ears are large, wide at the base, and should have a minimum 90-degree arc of curl, with firm cartilage extending from the base of the ear to at least one-third of its height. Breeders describe the degree of curl as straight, first-, second-, or third-degree curl. Almost without exception, show-quality American Curls are those with the maximum third-degree curl.

▲ *This American Curl Bicolour Tabby is a longhair. It has a fine, silky coat but is otherwise identical in every respect to the shorthair type.*

Breed Profile

Life expectancy:	13–15 years
Adult weight:	3–5kg (7–11lbs)
Average litter size:	3–5

Temperament: A quietly active, curious, affectionate cat. Enjoys family life, suitable for living in a flat, but enjoys safe access to the outdoors too. Responds well to training. Usually nonvocal but can occasionally be talkative. Longhair and shorthair varieties available. Both are relatively easy to groom as they have little undercoat.

Colours: Self, tortie, smoke, shaded, tipped, tabby, silver tabby, bicolour, tabby bicolour, pointed, lynx.

KNOWN HEALTH PROBLEMS

There are no special health problems of the American Curl except those associated with the deviation of the cartilage of the ear flap, or pinna.

Mismatching is the most obvious, where there is an obvious difference in the degree of curl in each ear. This is not generally a problem to the cat.

Calcification and skin disease can develop where the cartilage is kinked excessively.

Sunburn and Skin Cancer are a risk, especially in white American Curls, as the curl exposes more of the sensitive inner pinna to UV light.

● *Are the American Curl's ears easily injured?*

Pulling on the ears or uncurling them can cause damage to the cartilage, which is painful to the cat and can ruin a show cat.

● *The breeder insisted on keeping our American Curl kitten until it was 16 weeks old. Why was this?*

The ears begin to curl at a few days old but do not reach their full curl until the kitten is about 16 weeks old, so breeders tend to hang on to them in case they prove to be championship material. Prospective buyers should be sure that the breeder does not neglect the kitten's socialisation and that vaccinations have been carried out on schedule (see pages 18–19).

● *American Girl, our American Curl, is now 9 months old and in heat. We would love to have kittens from her. Is it a problem to mate her to another American Curl?*

No. The American Curl can be bred without fear of any problems, as many generations of Curls have been bred without any lethal gene appearing.

● *Do American Curls breed true? A friend's cat that was mated to another Curl produced some kittens with straight ears. Is she a crossbreed?*

Probably not. The Curl gene is dominant, which means that any cat carrying the gene will have curled ears. If a Curl heterozygote cat (which carries noncurl genes from one of its parents) is mated to another heterozygote cat, about 75 per cent of the litter will have curled ears and the remainder will be straight-eared. Both parents were still considered American Curl Cats. If, however, both parents have only the dominant curl gene (are homozygous), or only one is heterozygous, all the kittens will have the ear curl.

In general appearance the American Curl is a medium-sized cat with a semiforeign body and a modified wedge shape to its head. The muzzle is rounded with no break. The eyes are walnut-shaped with an oval upper lid rim and rounded lower rim. They are large, clear and bright, and may be any colour except in pointed Curls, which all have blue eyes. The shorthair coat is short and soft with a minimal undercoat lying flat along the body. The tail is wide at the base and tapers to be the same length as the body.

Curls are both alert and inquisitive, placid and adaptable. They are affectionate with their owners, and readily accept new situations and new people. In general they are not demanding or clingy, but they enjoy games and can be trained to do simple retrieving. Because of their origins as household cats, they love family life and being indoors. If you own an American Curl, you can expect to find it in your bed regularly, perhaps even on your bath. If allowed out, they are self-reliant and enjoy hunting small prey.

Ears are large and curl back. In spite of their appearance, they can swivel like ordinary cats' ears

Body is long and medium weight, with a smooth coat that may be short or long

Tail is long in proportion to the body and tapers towards the end

▶ *A black shorthair American Curl is a standard shorthair with a distinctive difference. It makes an excellent family pet as well as a show cat.*

Manx

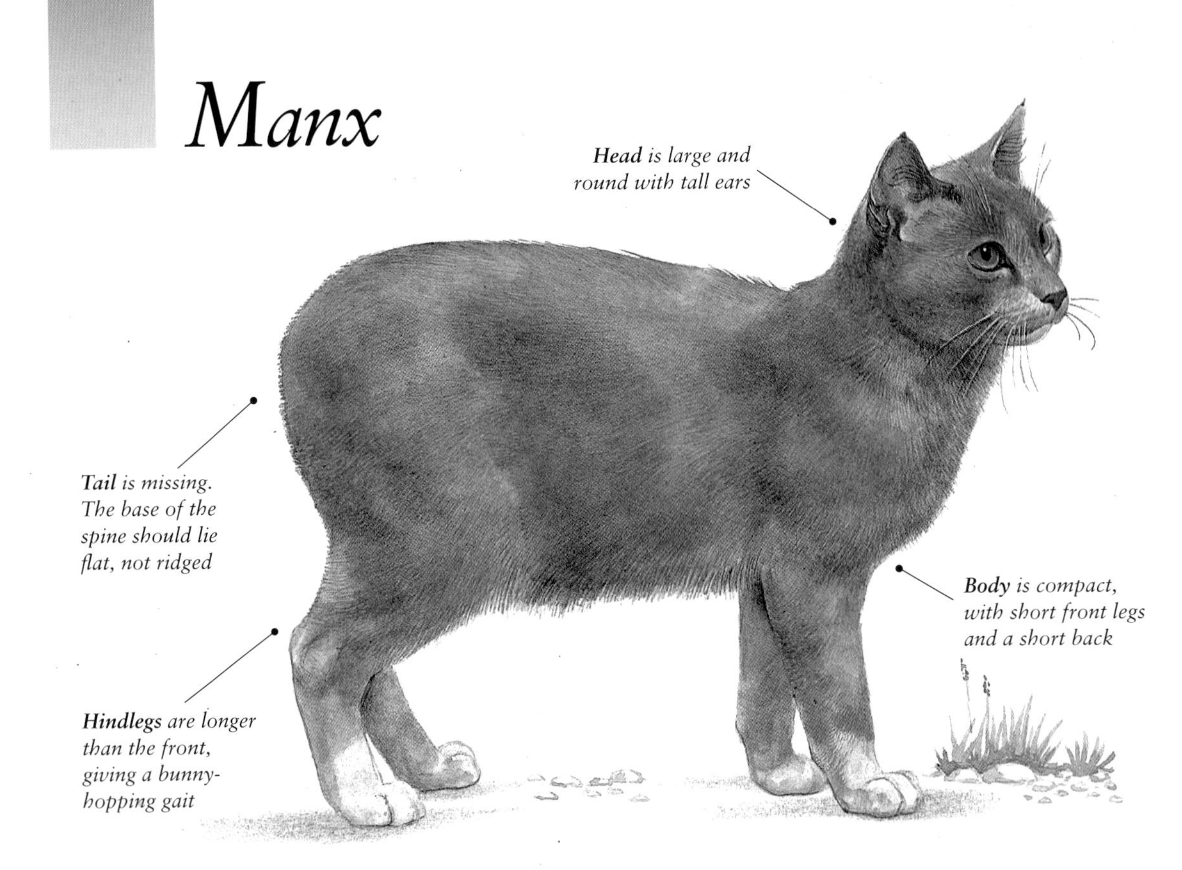

▲ *The Manx is round in face, cheeks, eyes, rump and body shape. The rear end sticks up even when the cat is standing with its weight evenly distributed. This blue-cream and white cat is a tail-less 'Rumpy'.*

THE TINY ISLE OF MAN IN THE IRISH SEA IS THE home of this unusual cat with no tail. One legendary explanation for this deformity is that the tail was trapped in the door as the first Manx squeezed aboard Noah's Ark. In fact, the lack of a tail is a genetic mutation that occurs occasionally in all animals. It usually disappears from the population in the course of normal breeding. In an isolated population, however, as found on an island, there is a much greater chance that such a gene will be perpetuated. A similar breed, the Japanese Bobtail, originated in the same fashion.

Rumpy, Stumpy, Longie or Riser

The genetic mutation is not expressed identically in all Manx cats, and there are four recognised varieties with residual tails of varying lengths. The only acceptable form for showing is the 'Rumpy', the most nearly tail-less Manx, which has merely a small dimple on the rump where the tail would normally be. The other extreme is the 'Longie', which has a shortened but otherwise almost normal tail. Between these two extremes are the 'Rumpy Riser', in which the tail is present only as a vestigial knob, and the 'Stumpy', which has a definite stump at the end.

The lack of a tail, combined with hindlegs that are longer than the front legs, causes a peculiar gait (described as bunny-hopping) that is characteristic of the Manx. Although the missing tail is clearly a deformity, it has no effect on the cats' mobility or agility, and they are usually excellent hunters. Selective breeding has created a very round face and body, often described as 'a hairy basketball with legs'. Manxes appear in many colours and patterns, but not all are recognised for show purposes in all countries. The double coat has a downy undercoat and a glossy outer coat of coarse hair. This can be groomed with a

● I have a Manx cat, Jilly, whom I wish to breed. I have heard it's not easy. Can you advise me?

If Jilly is a 'Rumpy' and you mate her to another Rumpy, there is a good chance that up to a quarter of the kittens will die in the womb or around the time of birth from congenital deformities due to Manx syndrome. If Jilly or the stud is another type, there is less risk, but if you are a novice, it would be sensible to find a more experienced breeder to provide the stud. Kittens with congenital defects rarely live out their third month and are usually euthanised to spare them suffering. Keep the kittens until they are 4 months old to ensure they are healthy before giving them away or selling them.

● We've been looking for a Manx kitten and were recently offered one. But when we went to see it, we were very surprised to see that it had a short tail. Does this mean it is a crossbreed?

Not necessarily, though Manxes are sometimes outcrossed with other breeds to minimise the chances of fatal spine problems. Your kitten could be a Longie, in which the genetic mutation of the spine produces a short residual tail. The kitten should have the affectionate, quiet Manx personality, and unless you are really set on a tail-less cat, there is no reason to refuse it.

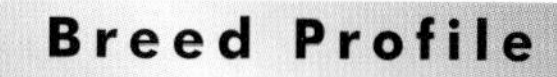

Breed Profile

Life expectancy:	13–15 years
Adult weight:	4–5.5 kg (9–12lbs)
Average litter size:	4

Temperament: Affectionate, quiet, intelligent cats. Not very active, but will enjoy being outdoors and are good hunters.

Colours: All shorthair colours; self, tortie, smoke, tipped, tabby, pointed and bicolour.

KNOWN HEALTH PROBLEMS

Manx Syndrome Due to the presence of the gene that foreshortens the spine, spina bifida (congenital deformity of the spinal cord) may occur, causing fatal malfunctions of the bowel or bladder. Pre- or neonatal death normally results.

comb or medium-stiff brush. In show cats, coat texture is more important than colour.

Manx cats have intelligent, affectionate dispositions. Although they are only of medium size, they are slow to mature. Manxes should be bred very carefully, especially two 'Rumpies', as the mutant tail gene often causes congenital spine problems, leading to a high fatality rate in litters and kittens up to 3 months old.

▼ The Manx is popular for its distinctive appearance, its amusing walk, and its relaxed temperament. Show cats are always 'Rumpies', but the other three types of Manx, which are not strictly tail-less, make good pets.

A Longhaired Manx

Early breeders noted that Manxes may produce longhaired kittens. This was the unintended result of cross-breeding, which introduced the recessive gene for long hair. During the 1960s a handful of breeders in North America decided to attempt to establish a longhaired Manx. They named it Cymric in honour of Wales, which was said to have its own variety of cats without tails. The Cymric possesses all the standard characteristics of the Manx, including its amiable personality, the potentially fatal gene that controls the development of the tail and its peculiar gait. In addition, the Cymric variety has a semi-long double coat and tufted ears and jowls. Unlike in the US, it has not yet been recognised by show associations in the United Kingdom.

Japanese Bobtail

ONE OF THE ANCIENT BREEDS, THE JAPANESE Bobtail is thought to have been imported from China or Korea. Written evidence and paintings show that cats have been present in Japan for at least a thousand years, but it is more difficult to pinpoint the beginning of a distinct breed with a bobbed tail. It probably originated from domesticated shorthaired cats that mutated into a tail-less version, as in the Manx; the gene was then perpetuated in the restricted gene pool of Japan's archipelago. Superstitious beliefs may also have favoured the development of the bobbed tail: the Japanese believed that a cat's tail with a clear split at the end was a sign of the devil, so short tails, with no obvious tip, were preferred.

Tradition holds that cats were the favoured pets of aristocrats, but in fact they were owned at all levels of society. In 1602 vermin threatened to ruin Japan's flourishing silk industry. All cats in the country were set free to keep down rodent numbers. They lived wild on farms or in colonies in the streets of Japan's early cities and mated at will. This effectively stopped the development of any pedigree lines, but in any case, all trade in cats had been banned.

The Discovery of the Bobtail

Japan was virtually a closed country for several hundred years, and it was not until after World War II that the Bobtail was discovered by the international cat fancying world. American military personnel stationed in Japan with the occupying forces organised the first cat shows in the country, but initially all native cats were ignored at the expense of imported breeds, even by the Japanese. Then, in 1963, American judges visiting a show in Japan spotted a Bobtail and were impressed. Five years later an American breeder in Japan sent some cats over to the United States, and when she herself eventually went home, she took 38 Japanese Bobtails with her. The breed has become established on the American show circuit but is not yet recognised in Europe.

Breed Profile

Life expectancy: 14–16 years

Adult weight: 2.5–4.5kg (6–10lbs)

Average litter size: 3–5

Temperament: Active, curious and friendly, both to people and other cats. An ideal family pet but may become destructive if bored; needs plenty of toys and activity. Suitable for flat living but should not be left alone indoors all day. Should be kept in pairs. Requires minimal grooming, even the longhaired variety.

Colours: All colours except pointed (Siamese) and agouti (Abyssinian).

KNOWN HEALTH PROBLEMS

There are no particular health problems associated with the Japanese Bobtail.

Cat with a Curly Bob

The distinguishing mark of the Japanese Bobtail is its short tail, 8–10cm (3–4in) in length, which is normally curled up to produce the bob. It can be straightened and is often held upright when the cat is alert or advertising its presence. Compared with the very rigid standards that apply to pedigreed cats from tail-less breeds such as the Manx, there is considerable freedom in breeding and showing Bobtails. Tail kinks and other traits usually considered flaws are permissible as long as the tail is distinctly present, fluffy and not more than 8cm (3in) long. Cross-breeding with the Manx has been strongly discouraged to keep the breeds separate and to prevent the congenital health problems that plague the Manx.

Selective breeding has progressively reduced the size of this cat since its discovery, as modern breeders prefer a more refined appearance than was seen in the free-living, nonpedigreed animal. The Bobtail today is medium sized, slim but well muscled, with a medium-short, silky coat and minimal undercoat. Typical of the 'foreign' type of shorthair, its head is pointed, with especially high cheekbones; these, coupled with the large, slanted eyes and long nose, give this cat a distinctive face that is often described as 'Japanese'. Eyes may be gold, blue, or odd.

Bobtails occur in all pedigree colours except pointed (Siamese) and agouti (Abyssinian), but the most prized variety is the Mi-Ke, meaning 'three fur'. This pattern consists of sparse but bold tortoiseshell markings against a pure-white background. The Mi-Ke is considered lucky in Japan, especially when accompanied by odd-coloured eyes. Also popular are black-and-white and red-and-white. There is a longhaired variety of Bobtail due to a recessive gene for a long coat, but this is still rare compared with the shorthair.

Bobtails enjoy swimming and can be taught to retrieve. They often have an endearing habit of raising the front paw as if in greeting. Friendly and curious, they make ideal family pets.

▼ *Japan's ancient good-luck cat is thriving in the United States. This cat is an example of the Mi-Ke ('three fur') coat. Most Bobtails are shorthaired, but there is a longhaired variety. Neither needs much grooming.*

● ***We love the look of the Japanese Bobtail and would like a female with a view to breeding her. What does this entail?***

Ideally, you should not attempt to breed a cat until she is 10–12 months old, and only then if you have time to devote to her in the late stages of pregnancy, during the birth and the first seven weeks of rearing the kittens. Get some initial ideas from your vet when you have her vaccinated, and locate a suitable mate well before you need him. You will have to take your cat to the stud cat for a day or two when she comes into heat. Japanese Bobtails always breed true because a recessive gene is responsible for the tail mutation.

● ***Turandot, my Japanese Bobtail, loves to swim, but I'm afraid the chemicals in the swimming pool will be bad for her. Should I keep her out of the water?***

Provided she has safe access and escape from the pool by solid steps, she should be quite safe swimming. However, the chemicals should be rinsed out of her coat immediately afterwards, just as you shower yourself. If her eyes become red after swimming, you would be wise to curtail the activity.

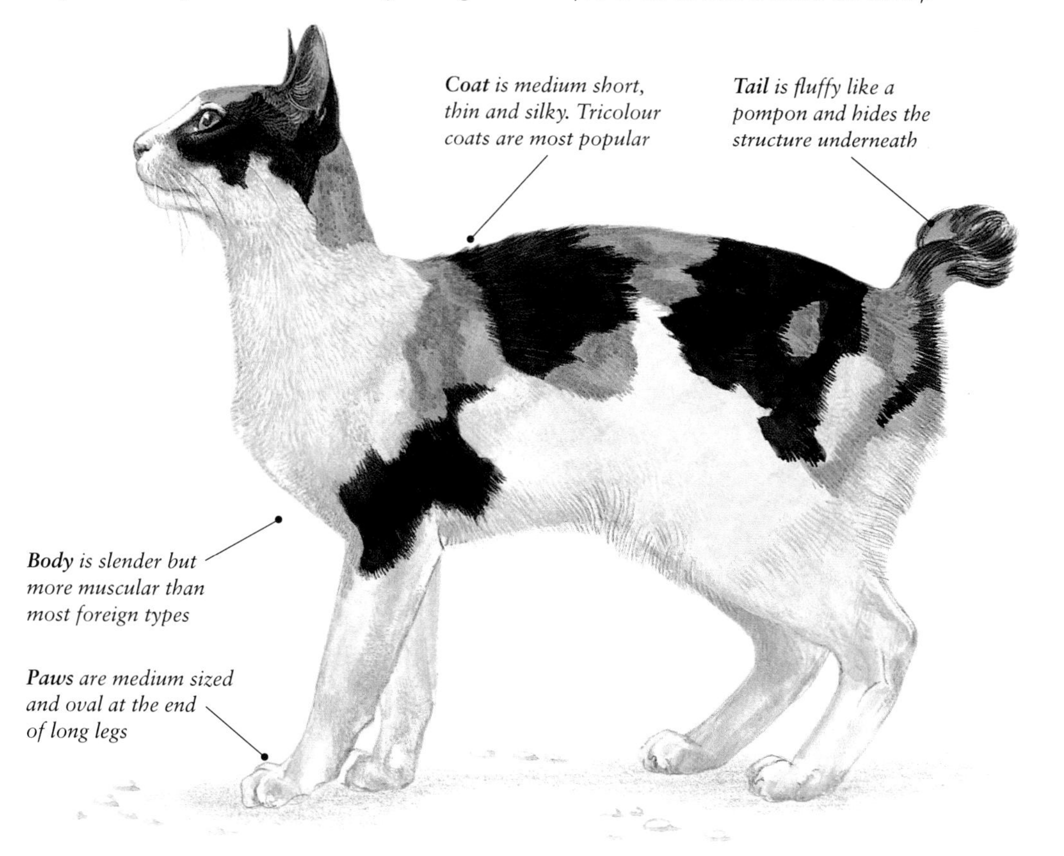

Coat *is medium short, thin and silky. Tricolour coats are most popular*

Tail *is fluffy like a pompon and hides the structure underneath*

Body *is slender but more muscular than most foreign types*

Paws *are medium sized and oval at the end of long legs*

Oriental Shorthair

THE EASIEST WAY TO THINK OF AN ORIENTAL is as a Siamese cat with a coat in solid colours. They share the same physical type, characterised by a slim, lithe body, wedge-shaped head and slanted eyes, and the same gregarious personality. Even the breed standards are indentical for each, except for colour. Siamese are pale with coloured points; Oriental Shorthairs may be any one of 50 different colours.

The intensity of the controversy over colour may surprise the cat owner who does not participate in shows. There have always been significant variations in the natural colour range of the Siamese, and it was only the uniqueness of the pointed version that made these cats the early

Breed Profile

Life expectancy:	16–18 years
Adult weight:	4–6kg (9–12lbs)
Average litter size:	6–8

Temperament: Curious, affectionate, demanding, but otherwise easy to own. A good beginner's cat: likes the activity of family life, playful. Suitable for living in a flat but an enthusiastic hunter if allowed outdoors. Appreciates company – it is better to keep a pair if you are are out during the day. Talkative but quieter than the Siamese. Easy to groom. Matures early and is easy to breed, producing large, active litters.

Colours: A wide range including black, brown, cinnamon, red, blue, lilac, fawn, cream, caramel, apricot, white; black, chocolate, cinnamon, blue, lilac, fawn and caramel tortie; smoke, shaded, tipped and tabby in self and tortie colours except white; silver tabby.

KNOWN HEALTH PROBLEMS

Cardiomyopathy (heart disease) is an occasional problem in Oriental Shorthairs.

Gingivitis (gum disease) occurs in the breed. Be especially attentive to your cat's dental hygiene.

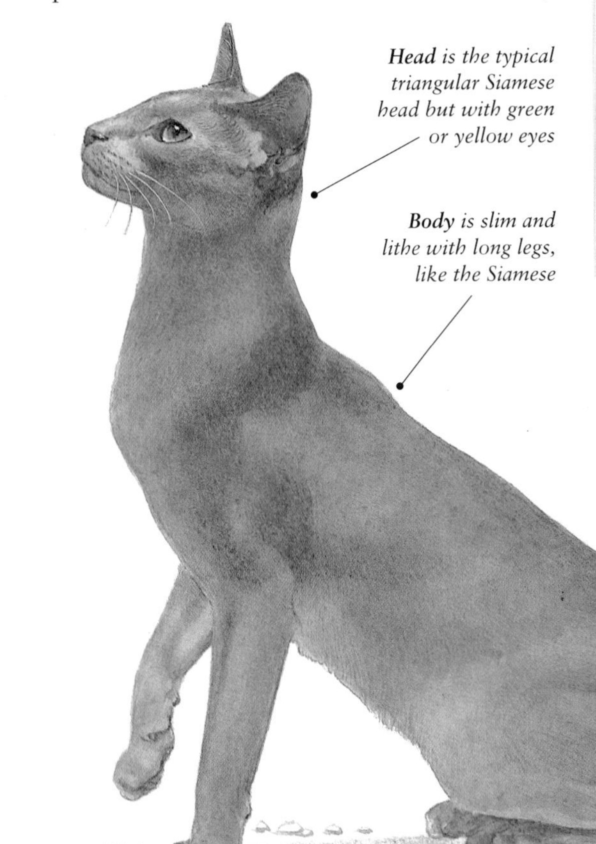

◀ *The Lilac or Lavender Oriental Shorthair was one of the earliest variations on the Havana and is still very popular. It was formerly known as the Foreign Lilac.*

favourites among pedigree cats. Nevertheless, a visitor to Thailand may observe that more than half the domestic cats are solid colours; only about 20 per cent of 'Siamese' cats possess the characteristic pale coat with points.

The two groups were split in the late 1920s, when the Siamese Club of Great Britain decided that only pointed cats with blue eyes could be registered as Siamese. The assortment of rejected types was large, but their numbers declined dramatically because these cats – designated merely 'Foreign' – had low status in the show world.

It was not until as late as the 1950s that breeders won recognition for a striking solid chocolate-coloured Siamese-type cat with green eyes, initially known as the Havana Brown. The breed was exported to the US under this name, but in Britain it was renamed the Chestnut Brown Foreign before reverting to Havana Brown. In the US the Havana Brown remains a separate breed. The popular Foreign Lilac and Foreign White Shorthairs were also named separately, but all the colours have now been classified as Oriental Shorthairs in an attempt to lessen considerable confusion among international cat fanciers. Blue Oriental Shorthairs are frequently confused with Korats (see pages 152–153), and Oriental Spotted Tabby Shorthairs with Egyptian Maus (see pages 178–179), but the Orientals look distinctly Siamese in their face and body shape.

Demanding, Not Delicate

Oriental Shorthairs are a very active, lively breed and want to be with their owner all or most of the time. They will join in any household activity and try to take over, whether you are reading a book or playing ball in the garden. It is important to provide toys or companions so that an Oriental Shorthair does not become bored, but as long as there is enough entertainment, these cats will live contentedly in a flat.

In spite of their delicate appearance, Oriental Shorthairs tend to live long lives. They are very prolific breeders and mature early, like Siamese, though they may be slightly less vocal. However, it is not wise to let them mate before 9 months of age. The typical Oriental Shorthair litter is large. Once the kittens reach 3–4 weeks of age, they are almost as active as their parents.

▲ *A tabby Oriental Shorthair is an extremely striking colour variation brought about by deliberate breeding. The large, almond-shaped eyes are very attractive.*

● Our white Oriental, Snowdrop, chews and swallows scraps of wool and cotton from clothes that we leave lying around. It's annoying, but will it hurt her ?

Yes – it can lead to bowel obstructions if left untreated. This behaviour is common in all Siamese types and tends to get steadily worse. See pages 44–45 and 66–67 for advice on training Snowdrop out of it.

● Our Oriental Shorthair, Su Mei, is 7 months old. During her first heat she made so much noise that the neighbours complained and we had to confine her for a few days. We want to breed from her when she is about a year old. Will we have to put up with this every time before then?

Su Mei will almost certainly keep coming into heat every week or so for a month or two. The only two ways to stop this if you wish to breed are to let her mate now, although it is a little early, or ask your vet for some temporary contraception. There are pills and shots that can stop the heat cycle until you wish to breed her. The timing of return to normal heat periods can be variable after hormone treatment.

● I am totally confused. My Oriental Blue Shorthair was mated to a Smoke Shorthair and produced two kittens that are clearly Blue Point Siamese. Do I register them as Siamese or Orientals?

It is confusing as rules vary from country to country, and even between cat fanciers' associations. Some will classify your kittens as Siamese, others as 'any other variety Orientals'. Some will not register them at all. Check with your local cat club.

Siamese

LEGENDARY GUARDIAN OF ROYAL TEMPLES IN Thailand (Siam), the Siamese is instantly identifiable, with its dark mask and vivid blue eyes set against a pale coat. The breed first appeared in Britain around 1870 and was promptly entered in a show in London. Its distinctive appearance, unlike that of any other breed then known, and its exotic origins guaranteed it popularity among wealthy cat lovers. The cats were known to be kept by royalty in their native country, adding to their mystique as glamorous pets for the elite. One pair in Britain came directly from the palace of the King of Siam; they were duly exhibited at shows soon after their arrival.

Siamese arrived in the United States by 1890 and caught on equally rapidly there, in spite of commanding prices that only the very rich could afford. Today they are rivalled only by the Persian and the Maine Coon in the United States.

Colour Controversies

Lithe and handsome, the Siamese has a short coat remarkable for its pale, almost white colour contrasted with a wide range of shaded 'points' (on the face, ears, paws and tail), and perhaps some shading on the flanks and back. This shading spreads over the coat as the cat grows older, so most Siamese are shown as young adults.

Most of the original Siamese exported from Thailand were the variety now called Seal Point, with the darkest black-brown points providing maximum contrast against the coat and eyes. Other colours appeared occasionally, but the Seal Point was the best known, and there was

▼ *Both the Tortie Point (left) and the Red Point (right) are relative newcomers to the breed. In most European countries they are classed as Siamese, but in the United States they are considered to be Colourpoint Shorthairs.*

▲ *Siamese kittens are born with light coats and acquire their colouring as they age. Because the breed is so high-strung and extroverted, it is important to give the kittens plenty of attention and amusement or they may become nervous.*

***Tail** is long, tapering and rather thin in proportion to the long, lithe body*

***Legs** are slender, with the hindlegs longer than the forelegs*

● My Seal Point Siamese, Natalya, has just had six lovely kittens by a Seal Point male. To my chagrin, they are all pure white. Did another cat get to her somehow?

No. All Siamese kittens are born white, and many owners are as confused as you are when the first litter arrives. The points will begin to colour after a week or two, usually beginning on the edge of the ears and the pads. You will also notice that their eyes, now pale, will begin to turn brilliant blue at about 8 weeks of age.

● I have just bought a female Blue Point kitten and intend to breed her with a Red Point. When should I mate her, and what colour kittens can I expect?

Siamese often start to call at about 5 months of age, but you should not let her mate until she is 10 to 12 months old and fully developed. The genetics of Siamese are complicated, and the colours do not all breed true. All your female kittens will be tortie points (tortoiseshell is a female sex-linked trait). The males will be whole colours at the points and may be seal, blue, lilac, or chocolate, depending on which colours the two parents carry but are not visible in their coats.

◀ *This Seal Point has the distinctive triangular head with large turned-out ears typical of the American Siamese. The highly curious, active disposition of the breed is reflected in its alert, intelligent expression.*

much resistance to accepting any other varieties as true Siamese; a Blue Point Siamese was disqualified from a British show in 1896 for being the incorrect colour. Blue Points and two additional colours – Chocolate Point (a diluted form of Seal) and Lilac Point (a frosty lavender-grey dilute of Blue) – were eventually recognised, and these four colours remain the only varieties accepted by the largest American cat associations. However, a larger number have been developed in the past 50 years. Most of these are recognised throughout Europe but officially classed as Colourpoint Shorthairs in the United States to maintain the distinction from Siamese. Still more confusing, self (solid) coloured Siamese have recently been developed; they are unmistakably of the Siamese type but lack points. They are known as Oriental Shorthairs (see pages 164–165).

The Problems of Popularity

Many early Siamese suffered from health disorders, such as gastric and respiratory problems. Their instant popularity created a huge demand that encouraged indiscriminate breeding and the widespread import of cats of questionable origins, so that by the 1950s the Siamese classes at cat shows were overcrowded, but some specimens were not sound. Their slanted, brilliant blue eyes tended to have a squint, causing some

Breed Profile

Life expectancy:	14–17 years
Adult weight:	2.5–4.5 kg (6–12lbs)
Average litter size:	5–7

Temperament: Very affectionate; should not be left all day without human or feline company. An ideal beginner's cat. Very noisy but easy to train. Suitable for families with or without children. Usually easy to breed. Will live in small flats, but particularly enjoy large houses and gardens as they are prolific hunters.

Colours: Seal, chocolate, blue and lilac point (the original colours); red, tortie, cream, cinnamon, fawn and caramel point; tortie and tabby variations; not all universally accepted.

Known Health Problems

Strabismus (squint) is an inherited defect of the Siamese. This defect is believed to compensate for their poor binocular vision, but although the cat looks cross-eyed, it seldom appears to suffer from any impairment of sight. There is no effective surgery or other treatment, and affected cats should not be bred from.

Tail Kink, a deviation of a tail bone up to 90 degrees, seems to occur more often in Siamese than other breeds. It is often found at the tip of the tail but can occur anywhere along the length. It is harmless but considered a showing fault.

Wool Eating is seen more often in Siamese than other cats. Any cloth left around the house is liable to have large holes chewed in it. This behaviour is thought to be stress-related (see pages 66–67, Stress and Phobias).

Mucopolysaccharidosis (MPS) is a rare developmental disease in Siamese, resulting in dwarfism, bone and joint disease, facial and tongue abnormalities and clouding of the cornea. Clinical signs start at 2–4 months of age, and affected cats usually die before they reach the age of 3 years. Bone marrow transplants have been successful in treating some cases.

Heart Disease is often hereditary in the Siamese, affecting the valves or heart muscle. Some cats will have no symptoms until later life, but others require medical treatment. Affected cats should not be used for breeding purposes.

cats to appear cross-eyed. Early breeders met with some success in eliminating the fault, but it reappeared in the 1940s and 1950s as a result of indiscriminate breeding. The tail sometimes features a kink at the tip, another fault in a show cat. Fads and fashion often influenced judges, so cats with more extreme traits were favoured for breeding. In this way the modern Siamese has gradually acquired an exaggerated angular line, especially along the head. A preference for a longer, tapering body shape has emerged in the United States, and the breed is dividing along these lines into yet more new classes.

Unlike most cats, Siamese can be taught to walk on a leash, to retrieve and to perform other tricks. This unusual feline trait is accompanied by an exceptionally outgoing personality. Siamese just love people and will await your return from work so that they can spend the rest of the evening perched on your lap. They should not be left alone for long during the day even if they have toys and a garden; if bored, they can become destructive.

Their affection does not necessarily extend to other cats. In spite of its delicate appearance, a Siamese is very capable of intimidating neighbouring cats if they invade its territory. They are also very vocal under most circumstances, and females in heat are known for making a loud yowling noise that many owners find hard to bear. Siamese of both sexes mature earlier than other breeds, so if you do not look out, your 6-month-old kitten may become a parent without your approval. They also appear to be more fertile than most breeds, producing large litters with ease.

◀ *These young Siamese are wiry bundles of energy. The fully grown cats may be quite small, weighing only 2.5kg (6lbs). This makes them suitable for city life, though they also enjoy outdoor life to the full, given the opportunity.*

Burmese

▲ *The American Burmese is less Oriental-looking than its European descendants. Both are graceful cats, possessing much of the Siamese beauty but less of the Siamese tendency to nervousness.*

THIS POPULAR BREED HAS A CONTROVERSIAL history. In 1930 a dark-brown hybrid Siamese female from Burma was imported into the United States and sold to a breeder. She was mated with another Siamese, producing several dark-brown males that were then mated back to their mother to consolidate the distinctive colour. The small gene pool continued to pose problems, and, as a result, one of the three principal American show councils suspended recognition of the Burmese from 1947 to 1956. As the breed has developed, it has lost the characteristic darker points on its limbs and the strident tendency of its Siamese forebears, as well as acquired a much wider range of coat colours.

Burmese are alert, intelligent, inquisitive and beautiful. Although lack of handling in the first few weeks of life can produce a shy cat, most Burmese are lively, more affectionate than nervy, and respond surprisingly well to training. They will play with you for hours, almost like a dog.

A Less Exaggerated Siamese

Burmese have characteristic, slightly elongated wedge-shaped heads, but not as exaggerated as the Siamese. The head is rounder in American Burmese, and less so in the European but with a slightly more rounded top. The body is surprisingly short and heavy for such an elegant animal, though again there is a transatlantic difference, with longer noses and bodies in the European breed. Both types have fine, sleek, glossy coats – a feature of the breed and a good indicator of general health. The first Burmese cats were a rich brown colour known as sable. Blue, champagne and platinum were subsequently recognised, and other colours are now becoming popular in the breed. Previously Burmese cats with these other

● *I've tried brushing, combing and bathing Hecate, my young Burmese female, but her coat never looks as glossy as I think it should. Could it be her diet?*

Follow your vet's advice for the best diet for Hecate, but this grooming tip may help: in addition to using a fine-toothed comb, try rubbing her with a damp chamois cloth. This cloth is widely used on show Burmese to help keep their coats in top condition.

● *Our Burmese, Brownie, is a champion wool eater. Is there anything we can do about this?*

Burmese are prone to eat any cloth you may leave around the house, from socks to towels. The condition is thought to be stress-related, but if your cat is basically happy – if he has enough company, enough toys, and access to the outdoors – the problem may have another cause. Try leaving small bowls of dry food around the house for him to nibble on. This will not eliminate the behaviour but should minimise it.

● *Do Burmese females in season make as much noise as Siamese, and do they mature as young?*

Burmese females may begin calling at about 7 months, which is slightly earlier than normal. They tend to be vocal in season but should not make as much racket as Siamese in this condition.

Breed Profile

Life expectancy:	15 or more years
Adult weight:	4kg (9lbs)
Average litter size:	5

Temperament: A highly affectionate family cat. Ideal for beginners. Vocal but not as much as Siamese. Can be trained to do tricks, especially retrieving. Suitable for living in a flat, but also enjoy access to outdoors as they are prolific hunters. Should not be left alone all day.

Colours: Sable (brown), blue, champagne (chocolate), platinum (lilac or frost), red, cream and various shades of tortoiseshell.

Known Health Problems

Health problems are rare, but there are some characteristic defects in the breed.

Tail Kink, an occasional deviation of a tail bone, often up to 90 degrees, is often found at the very tip of the tail but can occur anywhere along the length. It does not hurt the cat but is considered a showing fault.

Deformed Skull is an inherited deformity that first appeared in the 1970s. It is often fatal.

Flat Chestedness in kittens causes compression of the lungs and heart, leading to breathing difficulties and heart failure. It may be inherited and is usually noticed at 2–6 weeks of age.

▲ *A Blue Burmese mother hides her newborn kitten protectively under her chin. Burmese make excellent family pets, as they are highly affectionate. Owners prize them for their natural elegance.*

colours were known as Mandalays to distinguish them from the original Burmese. In all colours the underparts are lighter than the back; the ears and face are often darker too. Their large, rounded eyes may be any shade from chartreuse to amber, but most are deep gold.

Tiffanie: a Longhaired Burmese

Another variation on the Burmese is a longhair variety called a Tiffanie. The cross is sometimes made with a Chinchilla and sometimes with a self-coloured longhair. Tiffanies share the personality traits of the Burmese, and their large bright eyes are the same golden yellow. The coat, however, is long and finely textured, almost silky, and requires more dedicated grooming and combing than the Burmese. This is a fairly new breed that is still developing, and it may be difficult to find. Most Tiffanies are brown, but they may also be black, blue, chocolate, lilac, red, caramel, apricot, or any variety of tortoiseshell.

Tonkinese

THE TONKINESE IS A DIRECT BURMESE–SIAMESE cross and carries one gene from each breed, with neither gene dominating the other. The resulting kitten carries the features of both parents but in a unique modified form. Known popularly as the 'Tonk', the breed originated in Canada in the 1950s and was developed in the United States. The Tonkinese is now found all over the world and is recognised by most pedigree associations.

Reinventing a Breed

The Burmese itself began as a Siamese cross (see pages 170–171): Wong Mau, the dark-brown hybrid Siamese female who was used to found the Burmese line, has been described as the first Tonkinese. Some of her kittens would have been Tonkinese as well, and this fact is supported by the written records of the first breeder of the Burmese, who referred to 'light phase' kittens. These kittens were ignored initially while breeders concentrated on the Burmese.

However, both the Burmese and the Siamese were bred to extremes as their popularity grew. Two or three breeders decided to try to produce a cat that had the virtues of both the breeds as nature had first made them. Called Golden Siamese at first, the name was later changed to Tonkinese to emphasise the fact that this was a separate breed, not simply a hybrid Siamese. As Tonkinese cats began to be exhibited at cat shows, they scored an immediate hit with the public, though they initially inspired some rather mean-spirited comments from a handful of breeders of Burmese and Siamese who resented their success.

● ***My Tonkinese cat, Shadow, is now one year old, and I would like her to have kittens. Should I mate her to another Tonkinese, or to a Burmese or Siamese?***

The Tonkinese, because it is a hybrid, is interesting to breed. Although the results are predictable on average over a number of litters, litters usually contain a mixture of kittens. If you mate her to another Tonkinese, the resulting litter should contain two Tonkinese kittens for every one Siamese and one Burmese. If you mate her to a Burmese or a Siamese, half the kittens should be Tonkinese, and the other half will be of the breed to which you mated her.

● ***I have been offered a Tonkinese kitten, but I've heard that these cats are great wanderers. We live close to a very busy street. If I get a Tonk, should I keep it indoors?***

It would be unfair to keep such an active, adventurous cat indoors all the time unless you have a large house with room for a cat gymnasium and can give it lots of company. If you cannot guarantee a safe environment, you should consider getting a different breed.

▶ *Tonkinese kittens are born with pale coats and may take up to two years to develop the adult coloration. The Siamese points are more distinct in Natural (Brown) Tonkinese like these.*

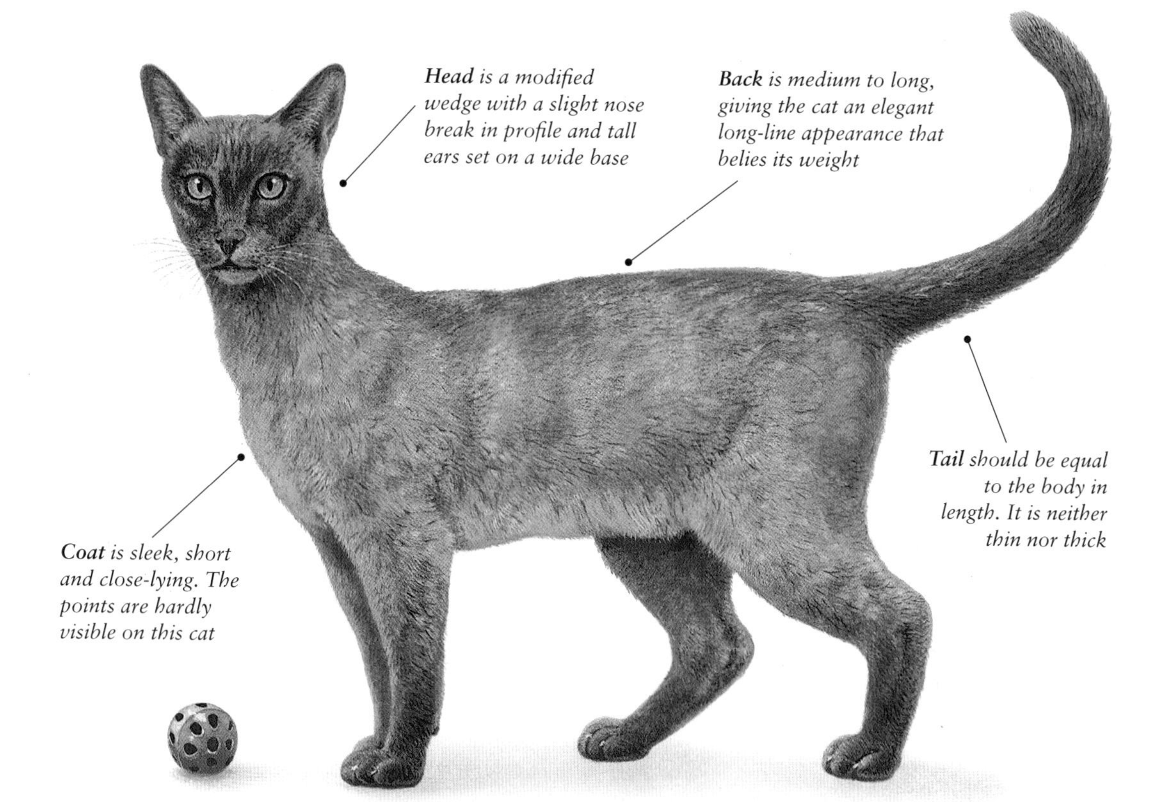

▲ *A Tonkinese is a hybrid cat that balances the two extremes of the pale, pointed, angular Siamese and the darker, rounder, heavier Burmese.*

Cat with a Mink Coat

The Tonkinese is a handsome cat with a short, silky coat and striking eyes, somewhere between blue and green, described as aqua. The Siamese points blend gently into the rest of the coat. The colours are often referred to as 'mink', a term that describes the fur texture as well. The most popular variety is the Natural or Brown Tonk that is produced from a Seal Point Siamese and a Brown Burmese. The American Champagne and Platinum Mink varieties are known as Chocolate and Lilac in Europe. All the colours seen in the Siamese and Burmese can be bred, producing a dilution of both colours.

Tonks are ideal family pets, as people-loving as Siamese but without the loud, insistent personality. They are also intelligent, playful, athletic and curious. They should not be confined to a small flat or left alone for hours every day. The coat requires only minimal grooming.

Breed Profile

Life expectancy:	12–15 years
Adult weight:	3–5.5kg (7–12lbs)
Average litter size:	5–6

Temperament: Sociable, affectionate, curious and active. A good beginner's cat. Suitable for families with children. Loves the outdoors. Not as noisy as the Siamese. Easy to train.

Colours: Mink: brown (natural), blue, champagne (chocolate), platinum (lilac); also all other Siamese and Burmese colours.

Known Health Problems

Special health problems of the Tonkinese can include problems found in Burmese and Siamese:

Squint (Strabismus) does not apparently affect the sight. There is no effective treatment. As it is inherited, affected cats should not be bred from.

Tail Kink A deviation of a tail bone of up to 90 degrees. It is painless but is a fault in show cats.

Deformed Skull is an inherited deformity in the Burmese line. It is usually fatal.

Flat-chested Kittens are born with their chests pressing against their heart and lungs, causing breathing difficulties and heart failure.

Abyssinian

▲ *A fawn Abyssinian, showing the breed's facial characteristics: a rounded, wedge-shaped head and a blunt muzzle. Adult males may have slight jowls.*

Breed Profile

Life expectancy:	14–16years
Adult weight:	4.5kg (10lbs)
Average litter size:	4

Temperament: Tends to be aloof with other cats but loves people. Good with children; can learn tricks. A good hunter. Talkative.

Colours: Ruddy, blue, sorrel, lilac, fawn and cream; also silver, silver sorrel and silver blue.

Known Health Problems

Progressive Retinal Atrophy (PRA), degeneration of the retina of the eye, occurs occasionally and can lead to total blindness.

Amyloidosis is a rare kidney condition caused by deposits of amyloid (a protein) in the tissues. Kidney failure usually follows. Affected cats or their relatives should not be bred from.

One of the oldest breeds, the Abyssinian has uncertain origins. Some admirers claim that it came from the Nile valley and was worshiped by the ancient Egyptians. An alternative view holds that Abyssinians were discovered in North Africa in the late 1860s by British soldiers who imported them to Europe. Still others insist that the Abyssinian is simply a British tabby cat with distinctive ticked fur. The breed was recognised in England in 1882 and is now popular all over the world. It is particularly strong in North America, where it was introduced in the early 1900s. American interest kept the breed alive during World War II, when the Abyssinian population in Britain declined sharply.

A Striking Appearance

In appearance, the Abyssinian is a striking cat that can be instantly identified by its patterned coat. It comes in a range of colours, but the most common form is golden-brown with black ticking (bands of dark colour along the hair), known as 'ruddy' in the US and as 'usual' elsewhere. The ticking is the result of a single mutant gene that is not found in other breeds and probably occurred thousands of years ago. It would have camouflaged the cat against the dry terrain of North Africa. Not all colours are accepted by all show registries, so if you plan to buy an Abyssinian for show purposes, find out which colours are accepted by your national association.

In physique, the Abyssinian is a medium-size cat of the exotic type with a lithe and muscular body, slender legs and small oval feet. Its coat should be short, close-lying and ticked. Its head is wedge-shaped but much less so than that of the Siamese, with its contours gently rounded. For show cats in the US, a slightly shorter wedge is preferred to that commonly seen in Europe. The ears are pricked, alert and wideset, and the eyes, which are also wideset, are typically deep amber, hazel, or green. The tip of the ear often has tufts of hair, a trait that is regarded as highly desirable in show cats.

An Ideal Family Cat

Apart from its exceptional beauty, the Abyssinian's popularity is due to its love of human company. It is highly intelligent, playful and quick to learn tricks. Although very talkative, it has a quieter voice than the other exotic types; even when in heat, it exhibits none of the raucous yowling typical of Siamese. It is usually gentle with children and makes an ideal family cat. However, the Abyssinian also loves its independence and prefers not to be part of a group of cats. It needs plenty of space in order to flourish, and for this reason it is more suited to country living than to an urban existence. Given the opportunity to be outdoors, it is a dedicated hunter and will bring home a steady stream of trophies.

A Longhaired Cousin

Closely related to the Abyssinian is the Somali, which differs only in its long coat – the result of a more varied gene pool in early breeding programmes. At first breeders insisted that the longhaired kittens that appeared from time to time in Abyssinian litters were not true Abyssinians. Now both varieties are well established genetically and the Somali, though still less popular, is a recognised breed in its own right.

● ***My 18-month-old Abyssinian, Sorrel, rips the furniture when I am at work. I live in a third-floor flat and can't let him out for the day. What can I do?***

Sorrel is clearly bored when you are out. Enrich his environment by leaving him toys to chase and boxes to explore – or get him a multigym. Teach him to use a scratching post instead of the furniture to condition his claws. If possible, ask someone to come and play with him during the day. It may be worth considering getting him a companion, but choose a breed that is suitable as an indoor cat.

● ***I took my Abyssinian, Leonie, to a cat show last week. She won third prize, but the judge said she was nervous when he examined her. She has seemed unsettled and less affectionate since then. Will she get used to being shown?***

She may, but Abyssinians do not like being confined, and the journey may always cause her some anxiety. Some cats love being shown, but these have usually been shown since they were kittens and are used to it. Unless you have your heart set on showing Leonie, it would probably be better not to continue.

▼ A Usual (or Ruddy) Abyssinian, showing the breed's body type. It is a muscular cat and an excellent hunter.

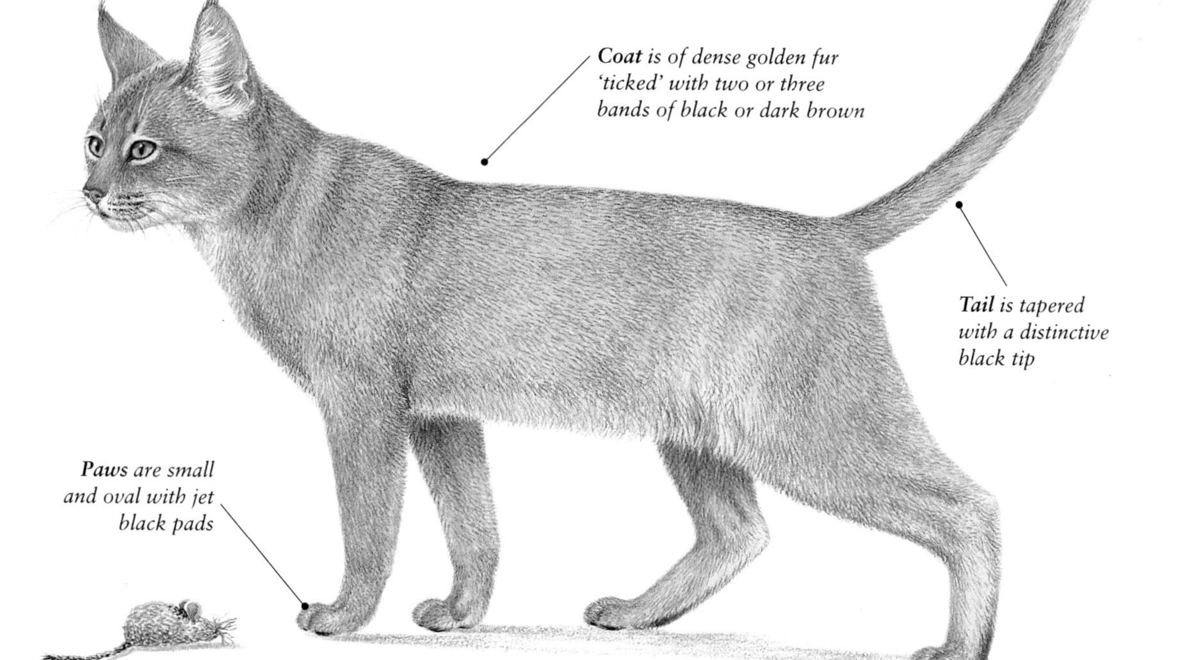

Ocicat

▲ *The Ocicat combines the exotic good looks and athletic grace of a wild cat with the people-loving disposition of a true family pet.*

IN THE 1960s AN AMERICAN BREEDER KNOWN for imaginative crosses created an Abyssinian-pointed Siamese by mating a hybrid Abyssinian-Siamese female to a Chocolate Point Siamese male. One of the offspring, when mated back to a Siamese, produced a litter containing Abyssinian-pointed Siamese – and a male kitten with unusual golden spots. At the time nobody thought twice about it, and it was sold as a pet for a token sum. However, its resemblance to a spotted wild cat, the ocelot, was remarked on by the breeder's daughter, who called the cat an Ocicat. Within two years this same cat had been noticed by a judge at an American show, and the breed was accepted for registration shortly after. However, its popularity did not take off until it obtained full championship status in 1987. Since then the registration of Ocicats has skyrocketed.

A Leopard on Your Lap

If you have always wanted to cuddle a leopard on your lap, at no risk to your life and limbs, this is the cat for you. The Ocicat is an imposing, wild-looking cat. With its sleek spotted coat and large size, it would be right at home in a jungle. There were only two original Ocicat colours, but more colours were introduced by crossing it with American Shorthairs. The full range of colours now includes all tabbies and silver tabbies.

The Ocicat is large, with a build somewhere between that of the short, sturdy Shorthair and the longer but muscular Abyssinian. It is much more powerful than its Siamese ancestors, and males are considerably larger than females. The short, wedge-shaped head has a broad muzzle with clear and definite facial markings, including the characteristic tabby M on the forehead and 'mascara' around the slightly slanted golden eyes. The ears flare out from the head. The short, fine coat should have clearly defined spots; these are less evident when the cat is moulting.

Ocicats look fierce, but they are gentle giants with outgoing, affectionate personalities. They love family life and should not be left without a companion (human, cat, or even dog) for long periods. The Ocicat's curious nature leads it to explore its surroundings, and it loves heights. If you do not want it to climb your shelves, curtains and kitchen cabinets, there should either be plenty of trees outside or a well-built multi-gym inside. Ocicats also love to entertain and will learn tricks such as retrieving.

● ***Our Ocicat kitten, Tonga, is already very large. Is he likely to outgrow a carrier if we buy it now?***

Yes. Buy an extra-large cat carrier or you will almost certainly have to replace it.

● ***We want to breed our tawny Ocicat female, Sheena, when she is a year old. Will she breed true if mated to a male Ocicat?***

Yes. She herself carries genes for at least the Siamese and Abyssinian, but if you mate her to another Ocicat, she will produce all Ocicat kittens.

● ***We have a Border Collie that's a real handful, and we'd like a cat. How about an Ocicat?***

An Ocicat would be ideal, especially if you get a kitten or a young cat. It is bigger and less temperamental than a Siamese, and it loves company enough to welcome the presence of even a boisterous dog. Your Ocicat may even be able to join in agility training.

Breed Profile

Life expectancy:	11–13 years
Adult weight:	3–7kg (7–15lbs)
Average litter size:	4–5

Temperament: Affectionate and devoted. Loves family life. Friendly even to strangers and dogs. Not suitable for life in a flat or long periods alone. Needs plenty of space, exercise and entertainment. Can be taught simple tricks.

Colours: Blue, chocolate, cinnamon, fawn, lavender, in self and silver; self silver; tawny (brown spotted or 'usual').

Known Health Problems

There are no special health problems associated with the Ocicat.

▼ *The accidental product of Abyssinian-Siamese crosses, the Ocicat is heavier than its athletic build suggests. It needs plenty of activity and space.*

Egyptian Mau

EGYPTIAN MAUS (LITERALLY, EGYPTIAN 'CATS') are a very ancient breed. Images of this striking spotted cat can be seen on tomb paintings in the Pyramids and on many other ancient Egyptian scrolls and paintings from as early as 1400 BC. The breed is so similar in appearance to African wildcats that many experts are convinced that the Maus are their direct descendants. They are even believed by some to be the earliest domesticated cats in the world, used by the Egyptians to prevent rodents from destroying the stores of grain that formed the basis of Egyptian civilisation. Spotted cats still live in the streets of Cairo today. Maus were probably introduced into Europe by Phoenician traders, beginning the line of European domestic shorthairs long before the Christian era.

▼ *The Egyptian Mau is a beautiful, adventurous cat that makes an excellent family pet. The markings show up particularly well against the silver coat.*

Three Egyptian Maus were imported into the United States in the 1950s by an exiled Russian princess – after narrowly missing disaster when they failed to get passage on a ship that sank in the North Atlantic – and all Egyptian Maus in the United States are descended from this stock. They were recognised for American shows in 1977, but the registered population is still quite small compared with that of other breeds, and the Egyptian Mau remains rare. Although they have been known across Europe for much longer than this, and competed in cat shows as early as the 1900s, they are still not recognised as a championship breed in the United Kingdom. The exotic splendour of the Mau's spotted coat makes it a common target for cat thieves. Make sure that your cat is tagged, preferably with a microchip, if you let it outdoors.

***Head** is a medium wedge with slightly large ears*

***Spots** along the back form a dorsal stripe*

***Legs** are medium long with stripes and well-developed muscles*

***Body** is less slender than that of most foreign-type cats*

Breed Profile

Life expectancy:	14–15 years
Adult weight:	2.5–5kg (5–11lbs)
Average litter size:	4

Temperament: Alert, intelligent and very affectionate, especially with its family – both feline and human. Good with children but does not like strangers. Vocal but quiet-voiced. Suitable for flats but will be happiest with lots of space for running, jumping, climbing and hunting. Prefers company and should be kept in pairs if the family is out for long periods and if they have no safe access to the outside. Easy to groom but needs grooming regularly.

Colours: Silver with pale silver undercoat and dark grey spots; bronze with copper undercoat, ivory underbelly, and dark brown spots; smoke (pewter) with pale silver coat tipped in black, silver undercoat, and black spots; black (no spots visible). Kittens are born with spotted coats.

Known Health Problems

There are no particular health problems associated with the Egyptian Mau.

● We live out in the country but have been told to keep our Egyptian Mau kitten indoors to keep her in top condition. Surely we could let her out?

Of course you could – and should, once she is old enough to be vaccinated and provided you do not live close to any busy streets or highways. Breeders and show enthusiasts tend to keep their Maus indoors, but if you do not intend to show your cat, there is no reason why you should not have her tagged and allow her outdoors. She will love it.

● I've heard that the Egyptian Mau is an outgoing cat, but the ones I've seen always look timid or worried. Is this a characteristic of the breed?

They do have a characteristic puzzled expression. This is emphasised by the facial markings, including ones that tend to make the cat look as if it is frowning. Egyptian Maus are inclined to be disdainful of strangers and may not be at their best when being examined by a potential owner. Breeders can prevent unfriendly habits from developing by handling kittens frequently from 2–3 weeks old. New owners should ensure that their new kitten meets as many people as possible very early on under relaxed conditions.

Short, Silky and Spotted

An Egyptian Mau's coat is short and silky. The markings on its body are random spots that vary in size and shape rather than following the more regular pattern typically seen in tabby cats. The spots form a contrast against the lighter background coat, which comes in four colours only. The bronze variety, a warm copper colour with pale sides and dark brown spots, bears the closest resemblance to the cat seen in the ancient Egyptian paintings. Some Maus are black, but are not acceptable for showing. Nose leather should be brick red in the silver and bronze varieties. The marks on the forehead form lines between the ears that continue down the back of the neck, breaking into oval spots along the spine. As these reach the lower back, they join together to form a dorsal stripe that runs along the tail to the very tip. The tail should be heavily banded and have a dark tip.

The Egyptian Mau is said to be the only naturally spotted cat, as distinct from other spotted cats, such as the Oriental Spotted Tabby, which have been developed through selective breeding. The latter breed is frequently confused with the Mau, but there are some significant differences: although it is unmistakably of the 'foreign' type (see pages 144–145), the Egyptian Mau is not as extreme in its physique, with a medium, muscular build, shorter legs and a head that is somewhere between round and wedge-shaped. The hindlegs are longer than the forelegs, so the cat appears to be standing on tiptoe. The Egyptian Mau is said to be the fastest domesticated cat, achieving speeds of up to 58kph (36 mph), and with its spotted coat it resembles a miniature cheetah. It is capable of leaping into the air to heights of 2m (6 ft).

With this kind of athletic prowess, this cat is unsurpassed as a hunter and is understandably happiest when it leads an active life. Its direct link to African wildcats probably accounts for these traits. However, after 40 years of careful breeding, little remains of the wildcat temperament. Although the Egyptian Mau may remain aloof with strangers, it makes an affectionate companion, devoted to its human family and good with children, and friendly to other cats. It has a quiet, melodious voice.

Longhaired Cats

Blue Persian

Longhaired coats developed as a genetic mutation of the short coat that occurs naturally in all cats in the wild. Because the gene that controls long hair is recessive, it may be hidden in shorthaired kittens from shorthaired parents who carry only one of the longhair genes, appearing when two such 'carrier' parents mate and produce a litter. Longhaired pedigree kittens from shorthaired parents were often shunned until a lone breeder appreciated the beauty of the type and began a campaign for its recognition. Some of the world's favourite longhaired breeds began this way.

The classic longhaired cat is the Persian, also known as the Longhair in many European countries. Typically it has a very long soft, silky coat consisting of guard hairs up to 12cm (4¾in), and numerous finer down hairs that are almost as long. This coat needs daily grooming.

There are many other breeds and crossbred cats that are longhaired. The Norwegian Forest Cat and Maine Coon have much shaggier and heavier coats than Persians. Some breeds are more properly called semi-longhairs, and most of their coat may be shed during warm weather, leaving the cat looking not much different from a shorthair.

Ragdoll (above)   Norwegian Forest Cat (right)

Persian

THE PERSIAN IS, ALMOST WITHOUT EXCEPTION, the most popular breed of cat in the world. For every other kind of pedigree cat in the United States, there are three registered Persians. They may have the longest history as champions, having appeared in shows for over 100 years.

All domestic cats native to Europe are shorthairs, and the Persian's most likely place of origin is in the mountains of Asia Minor. About 1620 both Persians and Angoras, admired for their long, silky coats (contrary to earlier belief, this was the result of a spontaneous genetic mutation, not of climate), were imported into France and Italy by members of the nobility. No distinction was made between the two types until a British publication of 1889. Persians had soon acquired an enthusiastic following in Britain, where they were originally known as French cats. Introduced at about this time into the United States, Persians rapidly became more popular than the local longhaired breed, the Maine Coon (see pages 194–197). Though they are classified as Persians in the United States today, their official name in Britain is Longhairs. Throughout the world, however, breeders still commonly refer to them as Persians.

▼ *Blue Persians were the first colour to appear after the traditional white and dominated shows for more than 50 years. Queen Victoria owned two.*

***Head** is big and round, with small ears and a short, broad nose*

***Tail** is short but very full to match the luxurious coat*

***Body** is massive and powerful, with a short neck, a broad chest and short, stocky legs*

***Paws** are large, round, and may be tufted*

▲ *All Persians are large, solidly built longhaired cats. Though there is wide variation in colour, there are few physical differences. Both these cats are tabby Persians (a red tabby above, a blue tabby below), which are comparatively rare, as the tabby pattern is difficult to breed in longhaired cats.*

Breed Profile

Life expectancy:	13–15 years
Adult weight:	4.5–7kg (9–15lbs)
Average litter size:	3–4 kittens

Temperament: Quiet and affectionate. Suitable for living in a flat but will also enjoy the outdoors. Long fine coat tangles easily and needs daily grooming to prevent matting.

Colours: Black, white, chocolate, blue, red, cream, lilac, tortie, chocolate tortie, blue cream, lilac cream (all self colours); silver tabby, all colours; also in smoke, shaded, tipped, tabby and bicolour.

Known Health Problems

Coat Matting occurs if daily grooming of the thick silky coat is neglected, especially around the base of the tail and abdomen. Hair that has come away from the hair follicle is held in place by new live hair, and this tangle becomes mixed with saliva as the cat tries to remove it. Professional help is often needed to remove these mats.

Epiphora (eye discharge) is common in flat-faced Persians, caused by partial or total obstruction of the tear ducts. Tears stream down the face from the inner corner of the eyes, leaving a persistent stain on some cats' fur. Careful bathing and cleaning of the eyes will help alleviate this (see Eye Problems, pages 98–101).

Progressive Retinal Atrophy (PRA) is a progressive degeneration of the retina of the eye that may lead to total blindness. It is thought to be hereditary. There is no treatment.

Deafness is seen in all breeds carrying the dominant gene for white coats, especially where the eyes are blue. There is no treatment.

Polycystic Kidney Disease is hereditary in Persians and can be screened.

Cryptorchidism (undescended testicles) is common in male Persians. The condition is hereditary, and affected cats should be neutered before they reach full maturity (see Male Cats, pages 74–75).

Apart from the name of the breed, there are differences in the colours and types recognised as Persians or Longhairs across the world. In Britain, every colour is considered to be a separate breed; however, national pedigree associations in the United States recognise nearly 50 varieties of Persian. The beautiful tipped Chinchilla, for example, is a breed in its own right in Britain but in the US is considered a variety of shaded Persian. Two exceptions to this are the chocolate and lilac self-colours, which are classified in the US as Himalayans or Kashmirs rather than Persians, while in Britain they are known as Chocolate and Lilac Longhairs. White Persians were once the majority, but the most overwhelmingly popular colour today is black.

A Luxurious Cat

Few cats are as impressive in appearance as a healthy, well-groomed, young adult Persian. The key to good looks is grooming, which is crucial in maintaining the luxurious silky coat that is the breed's chief distinguishing feature. The coat, which is long, flowing, thick and fine, requires intensive daily brushing and combing if it is not to turn into a mass of snarls and tangles. Moulting occurs throughout the year, and many owners prefer to keep their Persians' coats clipped if they are not being shown. The tail is relatively short but very bushy.

Under their coats, Persians are big cats, solid and balanced, with massive heads in proportion to their short and compact bodies. The wideset eyes are large, round, full and prominent in the flattened Persian face. Apart from the white Persians, which typically have odd-coloured or blue eyes, most self-coloured Persians have copper or deep orange eyes. Silvered Persians such as silver Chinchillas tend to have green eyes.

With their gentle, placid temperament, Persians are ideal housecats. They love people and seem to enjoy being with them. This sociability extends to other cats; Persians will tranquilly accept new arrivals into the household. They are not noisy or boisterous, and will live happily indoors, but they are also competent hunters if allowed access to the outdoors. Typical of large longhairs, Persians mature slowly and should not be bred before they arc 2 years old.

▲ *A silver Chinchilla Persian with her kittens. Persians are born with shorter coats that grow as they mature, reaching adult length at about 18 months.*

Peke-faced Persians

This breed is a true Persian that has been bred with an extremely flattened face so that it resembles that of a Pekingese dog. This characteristic is so exaggerated that the nose sits just below an indentation in the face between the eyes, and it tends to cause problems with blocked tear ducts. Peke-faced Persians are only recognised in the US and Canada as a separate breed for show purposes, and apart from their faces, the other standards are as for Persians. Peke-faced Persians are only bred in red and red tabby.

● *I love white Persians, but I would not want one that is deaf. How can I avoid this?*

Deafness in white Persians is an inherited condition. It tends to occur most in blue-eyed ones, so if you chose an orange-eyed white, this would reduce the chances considerably. You could also have an odd-eyed white, with one orange eye and one blue eye; the cat may be deaf only on the blue-eyed side.

● *I often see Persians at shows with tears running down their faces and brown stains on their cheeks. Is this a health problem?*

A certain amount of clear disharge from the eyes is normal for cats, especially flat-faced breeds such as Persians. As long as there is no infection and the fur around the eyes does not become matted, this is not a health problem. The condition is seen less among pet-quality Persians, which may have more of a nose or muzzle than the breed standard permits, and so are less prone to blocked tear ducts. Choose a kitten from a queen that does not have the problem, and ask about the stud too.

● *I have heard that Persians can have weak jaws. What does this mean?*

As a result of breeding programmes for the exaggerated flattened face, the lower jaw has become more and more narrow. This means that the chin is less prominent and the front of the lower jaw can be narrower than the upper jaw, resulting in a poorly-fitting, weak jaw. Breeders are trying to correct this.

Turkish Angora

THE ANGORA IS ONE OF THE OLDEST EUROPEAN cats, originating in Ankara in Turkey. In the 16th century it was introduced into France and then shortly afterwards into England, and, along with the Persian (see pages 182–185), is thought to be the likely source of the gene for long hair in the native domestic cats of Europe, which were originally all shorthaired. At the time, the two breeds were almost identical except in colour; Turkish Angoras occurred in many colours, while all Persians were blue. Both were prized for the novelty of their long silky fur, but during the late 19th century the Angora was eclipsed by the Persian. Like other longhaired non-Persians of the time, it declined until it became practically extinct outside its native country, where a breeding colony was kept in a zoo. It was not until after World War II that some cats were exported to the United States and Europe and the breed was revived.

Which Angora?

The Turkish Angora in the United States is not the same cat as the Angora in Europe. The modern American cats remain closest to their Turkish origins; in Europe other breeds were used as outcrosses to introduce the full range of colours. All Angoras were originally renowned for their elegant, silky, white coats, and some breeders

▼ *Unlike most longhaired cats, the Turkish Angora does not have a large body. Despite its long legs and body, it is only medium-sized, and its overall line should be lean and lithe. The full coat develops slowly and is shed for the summer months.*

▲ *Originally Turkish Angoras were found only in white. Today they can be almost any colour. This red and white variety is particularly attractive.*

Breed Profile

Life expectancy: 12 or more years

Adult weight: 2.5–4kg (5–9lbs)

Average litter size: 4–5

Temperament: Affectionate, playful, active. Suitable for families with or without children. Needs company and plenty of amusement.

Colours: Black, red, blue, cream, tortie, blue-cream and white self; smoke and shaded (except white); tabby, silver tabby and bicolour.

Known Health Problems

In the Turkish Angora they are few, but include:

Tail Kink, a deviation of one of the tail bones, sometimes up to 90 degree, is often found at the tip of the tail but can occur anywhere along the length.

Deafness often occurs in blue-eyed and odd-eyed white cats. It is genetic; affected cats should not be bred from.

insist that this is the only true Angora. Now almost any colour found in the longhair breeds is accepted for showing.

Because the coat is single rather than double, it is less likely to become matted. It is glossy with a silken sheen, but lies flat over most of the body except the tail and the neck, chin and 'britches' (on the hindlegs), where it tends to frill. This full coat may take two years or more to develop, and once in place, it moults heavily during the summer so that the cat appears to be a shorthair with a fluffy tail. The fur is not heavy enough to be spun into Angora sweaters, which are usually made from the fur of Angora goats and rabbits.

Playful and Graceful

The overall impression of the Angora is one of a cat with 'grace and flowing movement', as the breed standard states. As well as being elegant, it is an intelligent and playful cat. Although not as noisy as a Siamese, it is similarly active and outgoing, and should be provided with company most of the time. Some owners say that it loves to play to an audience, and certainly the sight of an Angora cat leaping up onto your shoulders in greeting reminds you how athletic and affectionate they are. They love games and will quickly learn simple tricks such as retrieving objects.

● ***Does Cloud, my Turkish Angora, need daily grooming in the summer, when most of her coat has been shed?***

No – provided you handgroom Cloud every day by stroking her, a weekly thorough combing should be quite sufficient until her winter coat begins to come through again.

● ***We have chosen an odd-eyed white kitten from a litter (he is now 5 weeks old), and the breeder tells us that there is a chance of deafness. Is there any way we can tell?***

The breeder should have some idea very soon, but there are tests you can perform. Separate the kitten from the litter and make a sudden loud noise when he is not looking. It should be obvious from the kitten's reaction whether he can hear or not. Deafness is not a severe disability as long as the cat can see – and provided you do not intend to breed from him.

● ***My Red Shaded Angora, Fatima, is a year old, and her coat is coming in nicely. However, after her summer moult her face suddenly looks very pale. Is this normal?***

Yes. The shaded coat is slightly tinged with silver, which looks lighter when the heavy winter coat is shed. Cats with pale coats are prone to sunburn, which can lead to skin cancer, so you should be careful about the amount of time your cat spends exposed to the sun.

Balinese

***Tail** is long and has a thick or moderate plume*

***Head** has the distinctive Siamese mask and eyes but is less triangular*

***Legs** are long, slim and elegant, with small feet*

***Coat** is fine, silky and lies close to the body. It is short in the front and longer at the back*

▲ *The Balinese is a graceful semi-longhaired cat with many of the features of the Siamese, whose genes it shares. It has the Siamese colouring, elegance and temperament, making it a loving family pet that needs a high level of mental and physical stimulation.*

DESPITE ITS NAME, THIS CAT DID NOT COME from the tropical island of Bali in Indonesia but was bred on purpose in the United States. Its graceful physique and balance inspired one of the original breeders to propose the name Balinese, in honor of the equally poised and elegant temple dancers of Bali. This was also a subtle way of suggesting that this breed, though very similar to the Siamese in appearance, was something separate. In fact, the Balinese is no more and no less than a longhaired Siamese, but the name of a breed can be a highly sensitive political issue in the world of pedigree cats.

Longhaired kittens appear from time to time in all shorthaired breeds and are usually not considered suitable for show. When this occurred in the Siamese in the United States in the 1940s, some breeders decided to develop them as a new breed. This was not difficult: because the gene for long hair is recessive, these longhaired cats bred true. In the 1960s they were recognised by a few American cat fanciers, but they were not officially recognised as a new breed by the influential American Cat Fanciers' Association until 1970. However, Balinese have successfully overcome their controversial beginnings and now enjoy great popularity throughout the world.

Not Quite a Longhair

Although the breed began as a longhair, the Balinese is now more properly described as a semi-longhair. This is the result of the widespread use of Siamese in selective breeding programmes. The coat is not thick and double like that of the Persian, but lies flat and close along the body, and the long hairs are confined to the posterior

● *My Balinese, Sambal, hates being groomed. How often should I put him through the ordeal?*

Luckily Balinese do not have the woolly undercoat that a lot of longhaired cats have, so he does not need intensive grooming. Try a soft brush twice a week, which he should accept. If he still objects, polish him gently with a chamois cloth or even hand brush by stroking him from head to tail.

● *How long should we wait before breeding from Dancer, our Blue Point Balinese? She is 7 months old and is bound to come in heat any time now.*

Ideally, you should wait until she is at least a year old. Until then you have two choices: either keep her indoors, away from other cats, or ask your vet for a contraceptive (see Female Cats, pages 76–77).

● *I have just been to see a cat advertised as a Seal Tortie Tabby Point Balinese. It does not look particularly Oriental. Can there be such a thing?*

There certainly can, and there is, though American cat registries will consider this cat a Javanese rather than a Balinese. The tortie-tabby combination is one of the newer and more controversial varieties (see text below). The uneven shading and points may confuse your eye, but the cat should have the typical brilliant blue eyes, graceful build and outgoing personality.

Breed Profile

Life expectancy:	12–15 years
Adult weight:	2.5–5kg (6–11lbs)
Average litter size:	3–4

Temperament: Affectionate, active and noisy. Suitable for families with or without children. Will live in a flat but prefers plenty of space for exercise. Should not be left alone for long periods without company or access to the outdoors. Can often be taught retrieving tricks.

Colours: US: Seal, chocolate, blue and lilac point only. Elsewhere: All Siamese colours.

Known Health Problems

Special health problems of the Balinese are few but include:

Strabismus (squint) may affect one or both eyes but does not apparently impair vision. There is no effective surgery or other treatment, but it is considered a fault in showing. As the condition is inherited, affected cats should not be used for breeding.

Tail Kink, a deviation of one of the tail bones, often up to 90 degrees, occurs fairly frequently in the Balinese (as in the Siamese). It is often found at the very tip of the tail but can occur anywhere along the length. It is of no concern to the cat but is not acceptable in a show cat.

of the cat: below the stomach and on the tail. From a distance a Balinese can look just like a Siamese with a thicker tail. The body is similarly lithe and long, with slim long legs and small feet. However, the Balinese looks like an old-style Siamese, without the extreme triangular head seen now; the standard to which it is bred conforms more closely to the shape of Siamese in the 1950s and 1960s, when the Balinese was being developed. The head is wedge-shaped, not round or pointed. The sides of the wedge give a straight profile, and the neck is long and elegant. In the dark full mask of the face are set the sapphire-blue eyes that distinguish the Siamese and most of its crosses. The mask is connected by tracings to the large, pricked ears, which follow the line of the wedge. They may also bear tufts. The tail is long and tapering but plumelike.

As with the Siamese, colour is a matter of controversy. In the United States only the four original Siamese colours – seal, blue, chocolate and lilac points – are recognised in the Balinese. The more recently developed colours seen in Oriental Shorthairs (see pages 164–165) also occur, but these cats are classed separately as Javanese. In Europe and Australasia, as for the Siamese, the wider range of colours is recognised as Balinese.

A Noisy Extrovert

Like its forebears, the Balinese is an unmistakably extroverted cat: active, inquisitive, talkative and demanding, with the typically loud Siamese voice. Friendly and very affectionate, it wants to be part of the family and will take the antics of children in its stride, but it should not be left to its own devices for long periods if you do not want to risk the destruction of your furnishings. To prevent boredom and moping, provide plenty of games and space for exercise. The Siamese type also means early sexual maturity, so be prepared; female Balinese may begin calling when they are as young as 6 months.

Birman

LEGEND HAS IT THAT THIS BEAUTIFUL SEMI-longhaired cat originated from the temples of Burma. First bred as a pedigree in France in the early 1900s, the Birman was much slower to catch on in the United Kingdom, where it was not recognised until the late 1960s, and the United States. The original Seal and Blue Points were crossbred with Siamese and Persians, and there are now as many as 20 acknowledged championship colours. Many are recent developments of the 1980s and 1990s, a particularly exciting time for Birman breeders. Birmans are extremely popular at shows around the world, and the competition tends to be strong.

Beautiful but Not Spoiled

Birmans have great charm as well as beauty. They are proud and intelligent, and love people. Birman kittens are particularly mischievous and their sense of fun stays with them into adulthood. At the same time, they are gentle and sensitive, and hate to be in trouble with their owner. Both kittens and adults tolerate other cats and dogs quite happily. Because they are so sociable, they adapt very quickly to a new home, but they have a tendency to pine if they are left alone at home for long periods of time.

The Birman's body is long and massive, with thickset legs and strong, short paws. Adult males can reach as much as 8kg (18 lb) in weight. The head is strong, broad and rounded, with full cheeks and a well-developed chin. When the cat reaches maturity, the breed's characteristic mask covers the face from nose to forehead, including the whisker pads. Tracings of lighter hair connect the mask to the ears, which are medium

▼A classic Seal Point Birman is a magnificent animal with a luxurious, silky, semi-long coat and a dark face mask setting off its deep blue eyes. The cat is rather large with a muscular frame.

***Head** is strong to match the physique*

***Neck ruff** is seen on adults*

***Tail** is full and coloured to match the points*

***White paws** stand out sharply against dark point colour on the legs*

Breed Profile

Life expectancy:	12–14 years
Adult weight:	4.5–8kg (10–18lbs)
Average litter size:	3–5

Temperament: Very social and affectionate. Talkative but not usually noisy. Adaptable to most households including those with other pets. Requires daily grooming but does not need much activity or exercise. An elegant indoor breed.

Colours: Seal, chocolate, red, blue, lilac, cream; all colours in self, tortie and tabby point.

Known Health Problems

Health problems are rare, but there are some characteristic defects.

Strabismus (squint), affecting one or both eyes, is thought to be a reaction to compensate for poor binocular vision, but the cat seldom appears to suffer any sight defect. There is no effective treatment. Affected cats should not be used for breeding purposes.

Dermoid Cyst, a developmental defect of eye tissue, appears as a lump on the surface of the eye, usually in the outer corner. It consists of fibrous tissue, often with hairs protruding from it, and can be uncomfortable for the cat. Surgery is usually successful in correcting the problem.

Protruding Sternum, the result of abnormal cartilage projections, can occur at either end of the sternum (breastbone). The condition is thought to be inherited.

Umbilical Hernia, a protrusion of abdominal fat through the muscle wall at the point where the umbilical cord was attached to the kitten, is thought to be hereditary but can occur by accident during pregnancy or birth, and vary in severity from small to large.

in size and wide-set. The coat should be long (although it is a semi-longhaired breed), silky and slightly curled on the stomach, with a full ruff around the neck. The eyes, which are almost round, are blue – the deeper the better. The medium-length bushy tail should be in proportion to the body. White feet are characteristic of the breed.

The Seal Point Birman, the earliest, is still considered most typical of the breed. The Tabby Point Birman, developed more recently, has vertical tracings on the forehead (known as 'frown marks'), striped legs and a ringed tail.

▲ *The red point colouring is one of the most recent developments in the breed. Most of the body is pale, with the colour more evident on the face and tail. Red Points may have tabby markings and freckled noses.*

● ***I'm thinking of getting a Birman. Are they easy to train?***

Birmans are very intelligent, but as with all felines, they are individual and independent. If training is started from a young age and a close bond is developed between cat and owner, with patience and time you may well be able to teach a Birman simple tricks.

● ***I have just bought a Birman kitten, Newman. He is 12 weeks old. How do I manage his coat as he gets older to stop it from matting?***

The coat will not get as long as a Persian's, but it can grow thick and will mat if neglected, particularly during a moult. Regular combing to prevent tangles and a brushing to remove loose hair is necessary for an adult. Introduce grooming on a daily basis by using a very soft baby brush. As Newman's coat gets longer and thicker, introduce the use of a comb as well. Finally, try to ensure that these sessions are a pleasant experience for Newman by rewarding him with a treat afterwards; if grooming turns into a battle of wills, you may come out the loser.

● ***Do Birmans always have blue eyes, or is that only the breed standard?***

All Birmans have blue eyes, but they vary considerably in depth of colour. For show purposes, the most desirable colour is sapphire blue.

Ragdoll

***Head** is large, with the face mask setting off the deep blue eyes*

***Tail** is bushy and long in proportion to the body*

***Body** is long and broad, with a full chest and muscular hindquarters*

***Paws** are large and round with tufts. The Ragdoll has longer gauntlets than the Birman*

▲ *A Ragdoll is a big teddy bear of a cat with silky fur and striking good looks; this one is a bicolour. It is ideal as an indoor pet, which has made it very popular.*

THE ULTIMATE GENTLE GIANT AMONG CATS, the Ragdoll is the world's largest domestic cat and a relatively new breed. It originated in the 1960s in California, but the very first steps toward creating it were somewhat confused. A mitted Seal Point Birman male mated with a longhaired white female – possibly a Persian, but not a pedigree cat – who, in due course, produced a litter of oversized, pointed kittens with semilong hair. Cross-breeding the first two generations of these kittens produced the first pedigree Ragdolls. The name was coined by the first breeder because of the tendency of these cats to go limp in people's arms. According to one story, this was the result of their pregnant mother being injured in a road accident shortly before giving birth to her kittens, but this explanation does not stand up to scientific scrutiny.

The Ragdoll has long been a favourite in the United States. Like the Maine Coon, it is now a successful export to Europe and Australasia.

A Bigger Birman

In appearance the Ragdoll at first glance resembles a Birman (see pages 190–191). It is a longhaired, pointed cat that may have white mittens. However, it is a bulkier cat than a Birman, and the coat is more properly described as semilong, being shorter around the head and longer towards the tail. The soft, silky texture makes it less prone to matting than the fur of many longhairs, and only moderate grooming is required.

Unlike other pointed breeds, pedigree Ragdolls have comparatively dark coats. The coat patterns recognised for showing include pointed, mitted and bicolour, but other colours are now being developed. Seal and Blue Point Ragdolls are the most common. Pointed cats should have well-defined masks and points, as in the Siamese.

Breed Profile

Life expectancy:	11–13 years
Adult weight:	4.5–10kg (12–20lbs)
Average litter size:	3–4

Temperament: Relaxed, gentle, good with children. Suitable for indoor living, even flats, and tends not to hunt if let out. Does not require much exercise. Slow to mature.

Colours: Pointed, mitted, and bicolour in seal, chocolate, blue and lilac. (This is a new breed, and other Birman colours may follow.)

Known Health Problems

There are no particular health problems associated with the Ragdoll, but because the cat is large, heavy and not very active, it may easily become obese. It must not be overfed.

● *I have heard that Ragdolls don't feel pain. Is this true?*

No, certainly not. This misconception appears to have arisen from the reported accident suffered by the first Ragdolls' mother just before she gave birth. They feel pain like any other cat but may be less communicative about it.

● ***We live in a flat in a busy urban area. We are looking for a cat but want to avoid getting one that will feel restricted by being confined indoors. A friend has suggested that a Ragdoll would be suitable. I'm surprised, because they are so large. What do you think?***

Despite their size Ragdolls are very tranquil and are ideal for a flat. However, they like company, so if you spend a lot of time away from home, it would be a kindness to consider having two Ragdolls.

● ***Would a Ragdoll get along with a dog? We have a Miniature Schnauzer and would like to get a cat.***

The Ragdoll's placid temperament makes it an ideal companion for a dog, provided it is brought up with one. Your Schnauzer may bother the new kitten for a week or two, but with vigilance and good handling (see pages 48–49) they should soon learn to live together.

● ***Why is the life expectancy of a Ragdoll relatively short compared to other cats?***

There are probably two reasons. One, they are very large, and this puts extra stress on their body systems. Two, they are not that active, which means they have a tendency to be overweight. This can shorten their lives.

Mitted cats, which most resemble the Birman, should have similar points with the addition of white 'mittens' and 'boots' on the feet. Bicolour cats have an inverted V of white on the mask, starting on the forehead and extending down to cover the nose, whisker pad and chin. This should not extend beyond the eyes.

Everything about a Ragdoll is big. They have medium to large heads, with a flat plane between the ears. The full cheeks taper to a rounded, well-developed muzzle and a firm chin with a level bite. The nose should have a gentle break and be of medium length. The ears are medium, with rounded tips. A Ragdoll's large oval eyes are always blue, the deeper blue the better. The body is large and muscular, giving an unmistakable impression of suppressed power. The legs are fairly heavily boned, with the hindlegs slightly longer than the front. The coat lies flat against the body and breaks as the cat moves. A mane of long hair on the sturdy neck frames the face, and the tail is long and bushy.

Like Siamese, Ragdoll kittens are born white and develop points in their first few weeks of life. The coat colour and pattern will not be fully present until the cat is between 2 and 3 years of age.

A Cat for Indoors

Ragdolls are renowned for being placid characters and make good indoor pets, even in small flats. They do not require much exercise and appear to be uninterested in hunting. This feature has been an important factor in their rapid rise to popularity in Australia, where the exotic wildlife of suburban areas is increasingly threatened by the activities of domestic cats that are allowed outdoors.

Their low level of physical activity would suggest to many that Ragdolls are somewhat sluggish. In fact, Ragdolls are alert and intelligent and respond well to training, such as learning to use a scratching post. They love family life and get along well with children. The Ragdoll is possibly the only breed of cat that enjoys being unceremoniously picked up and carried around, just like a big teddy bear, which is exactly how it appears to children. Make sure youngsters do not abuse their living 'toy', as Ragdolls usually have soft voices and tend not to protest much.

Maine Coon

▲ *America's most popular pet cat, the Maine Coon, has it all: exceptional beauty, hardiness and an easygoing, friendly temperament. It makes a wonderful pet but needs plenty of space and access to the outdoors.*

THIS ROBUST LONGHAIRED CAT IS NAMED FOR the American state of Maine and its native raccoon, which the cat resembles with its distinctive colouring and long bushy tail. Like most breeds, the Maine Coon has more than one farfetched legend attached to its origins. One is that the cats evolved from matings between raccoons and feral cats, which is biologically impossible. An equally romantic belief is that the first Maine Coons were sent over by Marie Antoinette, the ill-fated queen of France, during the French Revolution. A third notion is that the breed is simply a longhaired version of the American Shorthair. However, it is most likely that the Maine Coon came from early Angora or Norwegian Forest cats brought in by sailors and mated with local shorthairs. These cats certainly existed in other states but it was in Maine that they received their earliest promotion as a breed.

Regardless of their origin, Maine Coons have been well established for over a century in the United States, winning national shows as early as 1895. They were eclipsed by the introduction of Persians, but remained popular as farm cats. In the past 20 years they have undergone a sudden revival. They are now the number one breed in the United States and the number two breed in Britain, though they have not yet been recognised for showing in Britain.

Big and Handsome

Maine Coons are very large and strikingly handsome. Males can weigh as much as 9kg (20lbs), while females are typically smaller. Their impressive 'wild' appearance is due to their size and their long, shaggy, flowing coat, which has a full ruff of hair around the neck. Strictly speaking, Maine Coons are only semilonghaired: only the coat at the back is long; the coat at the front is considerably shorter. Because the Maine Coon's thick fur is able to resist moisture it protects the

▼ *Maine Coons are big cats with a thick, shaggy coat. The brown tabby colour, shown here, is considered one of the classic Maine Coon varieties; it is also known as 'Mackerel'. This breed's wide range of colours is one of many reasons for its renewed popularity.*

● ***I'm considering getting a Maine Coon but am a little put off by all that shaggy hair. Are they difficult to groom?***

No, they're surprisingly easy. The hairs are long and silky, not downy like a Persian's, and brush easily. Most Maine Coons actually seem to enjoy being groomed.

● ***I have just brought home a 10-week old Maine Coon. Because they grow slowly, is the timetable for sexual maturity and neutering any different?***

Although Maine Coons may not reach full size until they are 3 to 4 years old, it is unwise to expect sexual maturity to occur any later than it does in other breeds. You should wait until a female is 12–15 months old before mating her, but both sexes can be neutered from 5 months of age.

● ***My 4-year-old Maine Coon, Kennebunk, weighs 11kg (25lbs). Is he too heavy?***

He is right at the top of the range, and if he is a standard-sized neutered male, I suspect he would benefit from losing some weight. Ask your vet at his next checkup.

***Head** is large and relatively square, with powerful jaws*

***Tail** is long and bushy, with rings like a raccoon's*

***Neck** is thick and powerful but not short*

***Legs** are long and strong, with big, round paws*

Breed Profile

Life expectancy:	12–15 years
Adult weight:	Males 6–9 kg (13–20lbs) Females 4–5.5kg (9–12lb)
Average litter size:	1–4

Temperament: An affectionate, playful, but independent cat, able to learn tricks. Enjoys family life. Very large and hardy. Needs plenty of space and freedom – not suited to life in a flat. Easy to groom despite having a long coat.

Colours: Black, blue, cream, red, tortoiseshell, blue tortie, white self; smoke and shaded colours (except white); brown, red, blue, cream, tortie, and blue tortie tabby; silver tabby; bicolours.

Known Health Problems

Hip Dysplasia is an inherited deformity of the hip joint. The ball on the femur (thigh bone) is misshapen to some degree, and the socket on the pelvis is usually more shallow than normal. It can occur in one or both hips, and in severe cases the cat is very lame by the time it is an adult. Diagnosis is by X-ray. Severe cases need an operation to remove the affected area of bone.

Cardiomyopathy is a condition caused by malfunctioning heart muscle. It may be restricted to a minor malfunction of the heart valves, producing a murmur when the heart is listened to with a stethoscope, or in severe cases heart failure. Treatment with diuretic and cardiac medicines may help in the early stages, but severe cases of cardiomyopathy will eventually prove fatal.

cat against cold winters but remains pleasant to touch and stroke. Unlike the Persian's, the coat of the Maine Coon is unmistakably that of an outdoor cat. It can stand a little neglect without turning into a mass of knots and tangles, but minimal effort by the owner will ensure it is always in good condition.

In the early days of breeding programmes in the United States, the Maine Coon was often used as an outcross to build vigour into the Persian line. There was more similarity between the breeds in those days, whereas the modern Maine Coon is differently proportioned than the Persian in every aspect: it has a much longer body and legs and a very long, bushy tail. It is in every way a large, colourful, powerful cat. The breed it resembles most closely is the Norwegian Forest Cat (see pages 198–199). Maine Coons from show-quality bloodlines tend to have longer, larger and lighter bones than animals born of a long line of farm cats, and there has been a tendency to sparse coats and low bone density in some of the revived pedigree lines.

Maine Coons come in most colours except the Siamese pattern, lilac and chocolate. However, the best-known colour is the brown tabby pattern. It is this colouration and the characteristic ringed tail, resembling a raccoon's, that gave the breed its name. Nowadays the coat is seen in many self colours, tabby, tortoiseshell (calico) and bicolours. The eyes may be gold, green, copper, or blue, and mismatched eyes can occur.

A Well-balanced Personality

Most Maine Coons are well balanced between self-sufficiency and dependence on people. They love company but tend to sit on the chair next to you rather than right on your lap. They want to know what you are up to in the next room and will follow you to find out. But although they want to help you with whatever domestic task you are engaged in, they rarely demand attention. They vocalise by making a unique, quiet 'chirping' sound that many people consider very pleasant.

Unlike most cats Maine Coons enjoy simple retrieving tricks and can be taught to walk on a leash. They generally live to a ripe old age and remain easygoing and playful. Their confident personality ensures that they usually get along well with other cats as well as with dogs and children. This no-fuss temperament, combined with their physical robustness, seems to make them more popular with men than any other breed of cat.

A female Maine Coon, if mated, usually has only one litter of four kittens a year. These are all likely to have different coat colours and patterns. Due to their very substantial size, Maine Coons develop more slowly than most cats, and growth may continue until their fourth year.

▶ *The rich colours of a tortie tabby Maine Coon are the product of careful breeding. The characteristic red tone of the paler parts of the coat is known as 'rufousing' and should be emphasised in a pedigree Maine Coon. Coat colour may vary widely within a single litter.*

Norwegian Forest Cat
Siberian Forest Cat

Breed Profile

Life expectancy:	12–13 years
Adult weight:	3–9kg (7–20lbs)
Average litter size:	4–5

Temperament: Generally reserved but much less so with its family. Loves the outdoors and is very athletic – not suitable for urban living. Matures very slowly due to its large size. Double coat is easy to groom. Moults heavily once a year.

Colours: All colours except Siamese pattern

Known Health Problems

There are no particular health problems in the Norwegian or the Siberian Forest Cat, but spine and dental problems may arise as the cats are selectively bred for longer noses and bodies.

Fairy tales from the mid-1800s, based on old Norse myths, mention a longhaired, bushy-tailed cat that lived in the local forests. Although these cats were very real, and much prized for their hunting ability by innumerable generations of Norwegian farmers, it was not until the 1930s that they began to appear in shows. Organised breeding did not get under way until the 1970s and the breed did not reach Britain and the United States until the early 1980s. Today it remains rare outside Norway.

The Norwegian Forest Cat has a strong resemblance to the Maine Coon (see pages 194–197). Both are big, robust, longhaired, almost weatherproof cats that developed in a cold, harsh climate. However, there are some slight differences in the faces and bodies. The hindlegs of Norwegian Forest Cats are slightly longer than the front legs, which may account for their extraordinary climbing ability. The coat is double, while in the Maine Coon this feature is allowed but not specified in the breed standard. A mature adult should have a ruff and 'britches' (or breeches – long fur on the upper part of the hindlegs). Like the Maine Coon's, the Norwegian Forest's coat is relatively easy to keep tangle-free. Moulting is heavy in warm weather.

This breed is also well balanced between power and elegance in its appearance. The head is strong and triangular in shape, with a straight profile. The large eyes should not be too round, and the large, upright ears are tufted. The body is long and

◀ *The Norwegian Forest Cat has a natural look, reminding some people of a lynx. Despite its formidable size and its love of outdoor pursuits, this cat makes a gentle, amiable family pet.*

strong, and the feet also have tufts of hair between the toes. A long, very bushy tail reaches as far as the cat's neck when extended. The wide variety of colours reflects the long history of this 'natural' farm breed.

Norwegian Forest Cats make gentle pets, but they tend to be reserved with strangers and highly territorial with other cats. They love to hunt and are also reported to be skilled at fishing.

Siberian Forest Cat

Developed in a climate even more harsh than that of Norway, the Siberian Forest Cat has extra protection against the elements in the form of a slightly oily outer top coat, which makes it even more waterproof than other longhaired northern cats. The ancestors of this breed are thought to have been in northern Russia for at least a thousand years, but it did not become recognised in its native land until the 1980s. The first exports arrived in North America in 1990.

Very similar to the Maine Coon and Norwegian Forest Cat, the Siberian Forest Cat may be closer to its wild origins than the others, with a high proportion of tabby coats – possibly due to mating with wild cats. The most traditional colour is golden tabby, but selective breeding may change this. In North America the preferred type has developed rounder eyes than in Europe and Russia. As in wild cats, the hindlegs of the Siberian Forest Cat are slightly longer than the front legs, giving superior athletic ability.

● ***Our Norwegian Forest Cat, Odin, prefers sleeping in the outside shed rather than indoors in the house. Is this normal, or could he be unhappy at home?***

Norwegian Forests are tough, weatherproof cats, able to cope with harsh outdoor conditions. With his long coat, it is more than likely that Odin prefers the cooler temperature of your outside shed than your heated home. Provided it is dry and draughtproof, you have no cause to worry. I'm sure he'll soon come looking for shelter if the going gets really tough.

● ***We have looked everywhere for a Siberian Forest Cat since we saw one at a cat show, but they seem very difficult to find. It's equally difficult to find information on them. Any suggestions?***

Your vet's office should be able to put you in touch with the nearest pedigree cat association, and they may know of breeders nearby. However, the Siberian Forest cat is a very new breed, and there are not many around yet. If you have access, try the Internet. This is a useful source of information on all pedigree breeds.

▼ *The Siberian Forest Cat is the most wild-looking of the longhaired 'natural' breeds. Females are slightly smaller and more fine-boned than the males.*

***Eyes** are rounder than in the Norwegian Forest Cat*

***Legs** are slightly longer in the back than in front, as seen in wild cats. The back is slightly arched*

***Tail** is bushy but not especially long, with a rounded tip*

***Coat** is long and double, with an oily layer on the topcoat for waterproofing*

Glossary

Abscess A painful, pus-filled swelling, usually in or under the skin, following a bite or scratch.
Acute Describes a disease that is sudden in onset and typically runs a short course.
Agouti The colour between the stripes of a TABBY cat. In non-agouti cats there is no contrast, producing a solid (SELF) coat.
Anoestrus The period of inactivity in the female cat's reproductive cycle. See OESTRUS.
Autoimmune disease A destructive immunity developed against part of the cat's own body.
Awn hair Secondary short hair with a slightly thickened tip.

Benign Describes a TUMOUR that is not MALIGNANT, recurring, or spreading.
Bicolour A coat with patches of two colours, one of which is white.
Biopsy A sample of tissue taken from a living animal for diagnostic purposes.
Blue A coat colour ranging from grey to blue-grey.
Breed PUREBRED cats, more or less uniform in colour and structure, as produced and maintained by SELECTIVE BREEDING.
Breed standard A description of the ideal specimen in each BREED, set by the national CAT FANCY associations.

Calico A distinctive patterned coat, usually black and orange but sometimes blue and cream. The coat pattern is sex-linked and only seen in females. Also known as TORTOISESHELL.
Carpal pad The additional pad on a cat's foot, part way up the foreleg behind the CARPUS.
Carpus The equivalent of a human wrist joint, between the elbow and the toes, which acts like a hinge.
Castration Surgical removal of the testes of a male cat.
Cataract A permanent opacity in the lens of the eye that interferes with vision.
Cat fancy A general term for organised cat showing and breeding.
Catnip (catmint) A perennial plant (Nepeta cataria) whose smell provides mental stimulation for many (not all) cats.
Chronic Describes a disease that is gradual in onset and typically runs a long course.
Classic tabby A wide-striped or spotted coat pattern.
Cobby A short, compact body shape, typical of shorthaired cats, with a round head, broad shoulders and a short tail.
Colourpoint A variety of Persian cat with a Siamese (pointed) coat pattern. See POINTS.
Crossbreed A cat of mixed parentage – may be either two distinct BREEDS or types or a mixture of many breeds.

Dew claw The first toe, on the inside of the paw, not in contact with the ground. In most cats they are missing from the hind limbs.
Dilute A pale version of a coat colour; e.g., cream is a diluted colour of red.
Dominant gene A gene that is always expressed when present. See also RECESSIVE GENE.
Double coat A coat with both GUARD HAIRS and DOWN HAIRS.
Down hair Short, soft, curly secondary hair.

Ear mite A PARASITE found in the ears of cats, belonging to the spider family.
Elizabethan collar A conical collar fitted around the neck to prevent the cat from biting stitches, its body, etc., or scratching its head.
Endoscopy The procedure of investigating internal body symptoms using fibre optic equipment.
Exotic A breed produced by crossing Persians with British or American Shorthairs.

Feral A domestic breed or type that lives wild.
Foreign A lithe, fine-boned body shape as seen in Siamese cats.
Frost In the US, a term for LILAC coat colour.
Fur ball Shed hair that the cat swallows when grooming and licking itself, which mats and causes vomiting or an abdominal obstruction.

Gene pool The number of animals of a SPECIES, BREED, or type that are available for breeding.
Guard hairs The long, coarse hairs in a cat's coat.

Haw See THIRD EYELID.
Heat Usually refers to the 6–10 day period of the female's OESTRUS cycle when she is able to mate.
Hock A hinged joint in the hindleg; equivalent to the ankle between the toes and the STIFLE.
Hybrid A cat whose parents are of different breeds.

Immunodeficiency A condition in which an animal's immune system is not functioning properly; often caused by a virus.
Intact A cat that has not been NEUTERED.

Lavender Alternative term for LILAC.
Lilac Pale grey/pink coat colour.
Lynx A Siamese pattern coat with tabby POINTS.

Mackerel tabby A TABBY cat with a coat pattern of narrow stripes.

Malignant Describes a TUMOUR with the potential to spread through the body, leading to death.
Mascara lines Dark lines on the face, especially around the eyes.
Mask The dark areas of the face in Siamese POINTED coats.
Metabolic disorder Any abnormality of the chemical processes that normally keep the body in balance.
Metacarpal pad The large pad at the back of a cat's foot.
Mink A combination of POINTED and sepia patterned coat, as in Tonkinese.
Mittens Solid white paws, as seen on Birmans.

Neuter To sterilise a male or female cat by surgical removal of the testes or ovaries. See CASTRATION; SPAYING.
Nictitating membrane See THIRD EYELID.
Nose leather The hairless modified skin around the nostrils.

Odd-eyed Having an eye of one colour (e.g., blue) and the other of another colour (e.g., yellow).
Oestrus The period in the HEAT cycle when a female is fertile.
Oriental Alternative name for FOREIGN.
Outcrossing The use of a different breed when producing PEDIGREE cats by SELECTIVE BREEDING.

Parasite An animal that lives on and derives its food from another animal, usually to its detriment.
Pedigree The written record of a cat's ancestry.
Pointed A coat pattern in which the POINTS are different in colour (typically darker) than the rest of the coat.
Points The body extremities (face, ears, paws and tail); of contrasting color to the main coat color in Siamese-type cats.
Polydactyl Having an extra toe on the front and/or hind feet.
Prolapse Protrusion to the outside of an abdominal organ.
Purebred A cat whose parents belong to the same BREED and are themselves of unmixed descent.

Quarantine Isolation period for an animal that has or is suspected of having an infectious disease. Often imposed on animals imported from abroad as a precaution against RABIES and other diseases.
Queen A mature female cat that has not been NEUTERED and may be used for breeding.

Rabies In mammals, a fatal viral disease of the central nervous system passed on in saliva from an infected animal, usually by biting.
Recessive gene The opposite of DOMINANT: a gene that is only expressed when both parents carry it, thus passing on two copies of the gene to their offspring.
Register The recording of a PEDIGREE cat with a CAT FANCY.
Ruddy In the US, an Abyssinian coat colour, also known as 'usual'.
Ruff The hair around a cat's neck, especially on longhaired breeds.

Sable In the US, a brown coat colour, as seen on the Burmese.
Scent marking Depositing scent from various body secretions (e.g., by rubbing or spraying) as a message to other cats.
Scruff The thickened skin on the back of a cat's neck.
Seal The near black colour of the POINTS of the darkest Siamese type.
Selective breeding Planned matings between cats in order to enhance their breed or type.
Self colour A SOLID-coloured coat.
Silvering Hairs on a cat's coat that are TIPPED with silver.
Smoke Heavily TIPPED hairs, the ends much darker than the roots.
Socialisation The process by which a kitten becomes familiar with other animals, including humans.
Solid A coat that is of a single, unbroken colour.
Sorrel A light brown Abyssinian coat colour, sometimes called red.
Spay Surgical removal of a female's ovaries and (usually) uterus.
Species Animals with common characteristics that are capable of interbreeding.
Spraying Leaving small quantities of urine, usually on upright surfaces, to act as a territorial marker.
Stifle The equivalent of the human knee joint.
Stud An INTACT male cat that is kept specifically for breeding.

Tabby A cat with a striped, spotted, or blotched coat. See CLASSIC TABBY and MACKEREL TABBY.
Tabby point A cat with a coat that has both a TABBY pattern and Siamese-type POINTS.
Third eyelid An extra membrane in the corner of the cat's eye. If it is visible, the cat is ill or dehydrated.
Ticking Black flecks in a cat's coat, as seen in the Abyssinian.
Tipped coat A coat whose hairs are tipped with a different colour.
Tom Any INTACT male cat.
Topcoat The long GUARD HAIRS of a cat's coat.
Tortoiseshell (tortie) See CALICO.
Tumour Abnormal cell growth, either BENIGN or MALIGNANT.

Ulcer An open sore on an external or internal surface of the body.
Undercoat The shorter DOWN HAIRS and AWN HAIRS under the longer GUARD HAIRS or TOPCOAT. See DOUBLE COAT.

Vaccination Protection against infectious diseases by stimulating the immune system.

Wean To persuade kittens that are totally reliant on their mother's milk for sustenance to start taking solid food.

CONTRIBUTORS

Caroline Bower BVM&S MRCVS: You and Your Cat, Adopting a Homeless Cat, Choosing Your Kitten, House-training, Grooming, An Outdoor or Indoor Life and all of Part 2

John Bower BVSc MRCVS: Death of a Pet, Pregnancy and Birth, Common Parasites, Ear Problems, First Aid and Emergencies and all of Part 4 except Birman

Sally Cheetham: Mating, Raising Kittens and Birman

Adam Coulson BVMS CertVR MRCVS: Getting the Most from Your Vet, Cat Anatomy, Urinary Problems, Weight Loss, Increased Appetite, Increased Thirst, Lameness and Street Accidents

Philip Hunt BVSc MRCVS: Dental and Mouth Problems

Chris Morley BVSc MRCVS: Infectious Diseases

Hilary O'Dair BVetMed CertSAD CertSAM MRCVS: Your Kitten's Diet, Your Healthy Kitten, Feeding Your Adult Cat, Caring for Your Adult Cat, The Older Cat, Skin Problems (part), Digestive Problems (part), Sneezing, Respiratory Problems, Collapse, Seizures and Convulsions and Loss of Balance.

Stephen O'Shea MA VetMB CertVC MRCVS: Injuries from Fighting

Neil Slater BVSc MRCVS: Digestive Problems (part)

Alix Turnbull BSc BVMS MRCVS: Transporting Your Cat, Holiday Care, Male Cats and Female Cats

Kevin Watts BSc BVetMed CertVD MRCVS: Skin Problems (part) and Skin Lumps and Swellings

Nigel Bray MA VetMB MRCVS: Consultant Editor (USA)

The publishers would like to thank Carol Krzanowski of the Cat Fanciers' Association for data on top breeds.

PICTURES

ABBREVIATIONS

AOL	Andromeda Oxford Ltd	**C**	Cogis
BCL	Bruce Coleman Ltd	**IF**	Isabelle Français
JB	Jane Burton	**TSM**	The Stock Market, UK

1 JB; **2** TSM; **5** AOL; **7** JB/BCL; **8** Ulrike Schanz; **9** IF; **11** Wara/C; **12** Hans Reinhard/BCL; **13** JB/BCL; **14** JB; **15** AOL; **16** JB; **17** TSM; **18** John Daniels/Ardea London; **19** TSM; **20** AOL; **21** JB; **22** JB/BCL; **23** AOL; **24** Hans Reinhard/BCL; **25** Vidal/C; **26** JB/BCL; **27tl, 27tr** JB; **27b** AOL; **28** JB; **29** Hermeline/C; **30** Lanceau/C; **31** JB; **32-33, 33** Your Cat Magazine; **34** Lanceau/C; **35** Hermeline/C; **36** Spectrum Colour Library; **37** Hermeline/C; **39c** R.T. Willbie/Animal Photography; **39b** A. Bartel/TRIP; **41** Foto Natura/Frank Lane Picture Agency; **42** JB/BCL; **43t** TSM; **43b** JB; **45** JB/BCL; **47** Hermeline/C; **48** Hans Reinhard/BCL; **49** John Daniels/Ardea London; **50-51** TSM; **52** Paul Kaye/Sylvia Cordaiy Photo Library; **53** JB; **54** Varin/C; **55t** Kim Taylor/BCL; **55l, 55r** JB; **56** Lili/C; **57** Hans Reinhard/BCL; **58** JB/BCL; **59t** TSM; **59c** Your Cat Magazine; **60–61, 61** JB; **63** JB/BCL; **64** Monika Wegler; **65l** Ulrike Schanz; **65r** AOL; **66** Fagot/C; **67** IF; **69** Hans Reinhard/BCL; **71** TSM; **74, 75, 77, 79** JB/BCL; **80** JB; **81, 82-83** JB/BCL; **83** Labat/Lanceau/C; **84** JB; **85** JB/BCL; **86, 87, 89** JB; **90t** Sherley's Ltd; **90r, 92** JB; **94** JB/BCL; **95** JB; **96** Dr. K.L. Thoday/Royal School of Veterinary Studies, Edinburgh; **97** Ulrike Schanz; **98t, 98b, 99, 101t** JB; **101b** JB/BCL; **102** JB; **103** TSM; **104** JB; **105** Monika Smith/Sylvia Cordaiy Photo Library; **106** JB/BCL; **108** JB; **109** The Veterinary Hospital, Estover, Plymouth; **110** Dr. K.L. Thoday/ Royal School of Veterinary Studies, Edinburgh; **111** JB; **112** Français/C; **113** IF; **114** Kim Taylor/BCL; **115** TSM; **116** Sally Anne Thompson/Animal Photography; **117** Urolithiasis Laboratory, Inc.; **118l, 118r, 119t, 119b** JB; **120** Foto Natural/Frank Lane Picture Agency; **122** JB; **124** Labat/ Lanceau/C; **126** Sally Anne Thompson/Animal Photography; **127** TSM; **128** Dr. Mike Targett/Queen's Veterinary School, Cambridge; **129** JB; **130** Ulrike Schanz; **132l** The Veterinary Hospital, Estover, Plymouth; **132r, 134, 135, 136** JB; **139** Jean-Michel Labat/Ardea London; **140** IF; **141** Bernie/C; **143** Hans Reinhard/BCL; **145** Werner Layer/BCL; **147** IF; **149** Lanceau/C; **150** Werner Layer/BCL; **153** Adriano Bacchella/BCL; **154** John Daniels/ Ardea London; **158** IF; **161** Lanceau/C; **165, 167** IF; **168** TSM; **169** Sally Anne Thompson/Animal Photography; **171** JB/BCL; **172** JB; **174** Adriano Bacchella/BCL; **176** Jean-Michel Labat/Ardea London; **181** Ulrike Schanz; **182** Hans Reinhard/BCL; **184–185** John Daniels/Ardea London; **187** IF; **191** BCL; **194, 197, 198** Adriano Bacchella/BCL.
Equipment photography by Mark Mason Studios.

Artwork
Tim Hayward **73b, 156, 173, 180b, 192**; Richard Lewington **89**; Ruth Lindsay **72, 98, 155, 159**; Denys Ovenden **73tl, 175, 177, 178, 199**; Peter Warner **73tr, 144–152, 160–170, 180t, 183–190, 195.**

The publishers would like to thank Rosewood Pet Products and Money & Friend pet shop, Abingdon, for their assistance in this project.

Index

040–913